The Contemporary City

Series Editors
Richard Ronald, University of Amsterdam, Amsterdam, Noord-Holland,
The Netherlands
Emma Baker, University of Adelaide, Adelaide, Australia

In recent decades, cities have been variously impacted by neoliberalism, economic crises, climate change, industrialization and post-industrialization, and widening inequalities. So what is it like to live in these contemporary cities? What are the key drivers shaping cities and neighbourhoods? To what extent are people being bound together or driven apart? How do these factors vary cross-culturally and cross nationally? This book series aims to explore the various aspects of the contemporary urban experience from a firmly interdisciplinary and international perspective. With editors based in Amsterdam and Adelaide, the series is drawn on an axis between old and new cities in the West and East.

NOW INDEXED ON SCOPUS!

More information about this series at
http://www.palgrave.com/gp/series/14446

Valentina Carraro

Jerusalem Online

Critical Cartography for the Digital Age

Valentina Carraro
University of Amsterdam
Amsterdam, The Netherlands

ISSN 2634-5463   ISSN 2634-5471   (electronic)
The Contemporary City
ISBN 978-981-16-3313-3   ISBN 978-981-16-3314-0   (eBook)
https://doi.org/10.1007/978-981-16-3314-0

Cover illustration: © Melisa Hasan

This Palgrave Macmillan imprint is published by the registered company Springer Nature Singapore Pte Ltd.
The registered company address is: 152 Beach Road, #21-01/04 Gateway East, Singapore 189721, Singapore

ACKNOWLEDGMENTS

I am grateful to the many people and institutions, in many countries, who helped me complete this research and write this book. In Hong Kong, I thank the Hong Kong Research Grant Council, which funded this project through its Hong Kong Ph.D. Fellowship Scheme, and the CityU School of Graduate Studies, for its financial and administrative support. Special thanks go to my Ph.D. supervisor, Bart Wissink, for guiding me through the process with dedication and empathy, for pushing me to put up with Latour's dudeish attitudes, and for his general supportiveness. I am also indebted to Ray Forrest, for engaging my research with all the curiosity and kindness a junior scholar could hope for, and then encouraging me to turn it into a book. It is to his memory that I would like to dedicate this book.

In Jerusalem, I am grateful to the many individuals who, with their warmth, insightfulness and sincerity, have left a mark on this book and its author. This includes, of course, the wonderful Grassroots Jerusalem staff, with a special mention to Amany and Fayrouz, who patiently put up with the questions of yet another well-intentioned ajnabi without a clue, and then extended their friendship to me. I am also thankful to the Israel National Library and the Kenyon Institute for providing access to books, maps and working space during my stay.

I am forever grateful to my family, particularly my parents: for their encouragement, trust, unwavering support and for making me the person who I am. Finally, there is the woman who lived through this with me, day after day: thank you, Tascha, for forwarding that one application call for a mapping post in Jerusalem, for enthusiastically supporting the plan to move to Hong Kong, and then to Chile, for keeping me sane during five months of lockdown and related vitamin D deficiency, and for still being up for more.

CONTENTS

1   New Mapping Technologies, Same Old Politics?   1
    'They Had the Plans, the Territory, the Maps…': Cartography
    and the Struggle Over Israel/Palestine (1923–2000')   3
    The Geoweb: A New Mapping Language   13
    Book Overview   18
    References   19

2   A Critical Cartography of Sensors and Algorithms   23
    Mappings: From Big Narrative to Big Data?   28
    Approaching Maps as Matters of Care   32
    References   36

3   Into Dangerous Territory with Waze   43
    Danger-Tracking Apps to Keep (White, Middle-class) People
    Safe   45
    When 'They' Are 'Here': Navigating Jerusalem's Soft Borders   49
    Palestinian Lives in Danger   55
    References   59

4   A Glitch in Google Maps   65
    The Google Maps Machine   67
    Erratum: *The Palestinian Question, in 140 Characters*   76
    References   83

5 Naming Jerusalem on OpenStreetMap 87
The Edit War Over Jerusalem 89
Names on the Ground: A Historical Detour 96
References 106

6 Epilogue 111
Gaps in the Digital Map 112
No Escape in Technology 118
References 121

Appendix: Notes on Methods 125

Index 129

Abbreviations

BDS   Boycott, Divestment, Sanctions
DWG   Data Working Group (OSM)
GIS   Geographic Information Systems
GJ   Grassroots Jerusalem
NGO   Non-Governmental Organisation
OSM   OpenStreetMap
PA   Palestinian Authority
PLO   Palestinian Liberation Organisation
RC   Refugee Camp
UN   United Nations
WADA   Waze Avoid Dangerous Areas

LIST OF FIGURES

Fig. 1.1   The neighbourhood of Ras al-Amud on Google Maps, as of October 2018 (*Source* Google Maps; Google and the Google logo are registered trademarks of Google LLC, used with permission)   2

Fig. 1.2   The 1947 UN partition plan (*Source* Map elaborated by the author, based on UN sources)   6

Fig. 1.3   The 1949 armistice lines (*Source* Map elaborated by the author, based on UN sources)   8

Fig. 1.4   The 1949 armistice lines in Jerusalem (*Source* Map elaborated by the author, based on UN sources)   9

Fig. 1.5   Israel's borders at the end of the 1967 war (*Source* Map elaborated by the author based on UN map 3243)   10

Fig. 1.6   Oslo areas (*Source* Map elaborated by the author based on UN sources)   12

Fig. 2.1   Map of the Jerusalem Light Railway (*Source* Map elaborated by the author using data from OSM [licensed under ODbL] and Peace Now [licensed under CC BY-IGO])   24

Fig. 2.2   The JLR train in 2017, decorated with the Israeli flag to celebrate the anniversary of the 1967 war and consequent annexation of East Jerusalem (*Source* Photo taken by the author)   26

Fig. 2.3   Mural of the Jerusalem Light Rail on Jaffa Road (*Source* Photo from Flickr, taken by user Paulina Zet _ Vered Hasharon [licensed under CC BY 2.0])   27

Fig. 3.1   Location of the Qalandiya refugee camp in relation
to the Jerusalem municipal boundary and West Bank
Barrier (*Source* Map elaborated by the author using data
from OSM [licensed under ODbL])   44

Fig. 3.2   Diagram illustrating the basic working of WADA
and similar navigation systems. The app calculates
the shortest route from start to destination that avoids
crossing the specified polygons. In the example,
the only viable option is route c   46

Fig. 3.3   Ras Al-Amud and Shu'fat. The map shows the position
of these neighbourhoods in relation to Jerusalem's
'hard' boundaries: the 1949 armistice line, the West
Bank barrier and the municipal boundary (*Source* Map
elaborated by the author using data from OSM [licensed
under ODbL] and Peace Now [licensed under CC
BY-IGO])   53

Fig. 4.1   Google Maps search results for 'Jerusalem',
as of September 2018 (*Source* Google Maps, annotations
in red by the author; Google and the Google logo
are registered trademarks of Google LLC, used
with permission)   68

Fig. 4.2   Comparison between the international and Indian
versions of Google Maps, as of September 2018. The
screenshot focuses on the regions of Jammu and Kashmir,
where China, Pakistan and India have conflicting territorial
claims. When search settings are set to 'global', Google
Maps shows the borders as disputed; when search settings
are set to 'India', the entire area is represented as part
of India (*Source* Google Maps; Google and the Google
logo are registered trademarks of Google LLC, used
with permission)   73

Fig. 4.3   Google Maps search results for 'Palestine', as of September
2018 (*Source* Google Maps; Google and the Google
logo are registered trademarks of Google LLC, used
with permission)   73

Fig. 4.4   Google Street View imagery of Jericho Road,
as of September 2018. The bottom half of the screen
shows the same road in Google Maps view. The West
Bank barrier, which blocks the road, is not reported
on the map (*Source* Google Maps; Google and the Google
logo are registered trademarks of Google LLC, used
with permission)   74

Fig. 5.1   Detail from Sandreczki. 1883. 'Plan Der Strassen &
Plätze Des Jetzigen Jerusalem'. Zeitschrift des Deutschen
Paelestina Vereins, Vol. VI [1883]. Related text: pp. 43–77.
Public domain   100

Fig. 5.2   British Mandate street sign. The name 'Street of Prophets'
is translated in three languages, English, Arabic
and Hebrew (*Source* Photographed by user DMY in 2007,
Wikipedia Commons [licensed under CC BY 3.0])   101

Fig. 5.3   A street sign in the Old City's Muslim Quarter.
The sign illustrates the struggle for linguistic
dominance over Jerusalem. Its bottom portion, put
up after the Jordanian administration came into power,
gives the Arabic name above the English transliteration.
The Hebrew name was added above the other names
after 1967 (*Source* Photographed by user Yoav Dothan
in 2009, Wikimedia Commons, Public Domain)   102

CHAPTER 1

New Mapping Technologies, Same Old Politics?

Abstract An introduction to the recent history of Jerusalem, underscoring the role of cartography. Here, I explore how maps have been used to support the Zionist project, both symbolically (by presenting Palestine as a land without a people) and materially (by facilitating military expansion and civic administration). I also highlight key technological and social transformations undergone by cartography in recent decades. While it would be naïve to think new technologies make maps any less political, my argument is that—in Jerusalem as elsewhere—there is a need to investigate their characteristics and effects. Do web maps 'repackage' the same substance in a new form, or do they bring about genuine change, for better (more democratic) or for worse (less transparent)?

Keywords Jerusalem · Palestine · Cartography · Geoweb · Zionism

Walking to my Jerusalem office, I soon learnt not to rely on maps. My journey started in Ras al-Amud, a lively neighbourhood in the East of the city that appears on Google Maps only as a grey expanse crossed by blank streets (Fig. 1.1). My phone screen would become more and more populated as I neared the city centre, but street names would often not match those reported on road signs, let alone those used by locals. The office was located on what had until recently been Ismail Hijazi Street,

© The Author(s), under exclusive license to Springer Nature
Singapore Pte Ltd. 2021
V. Carraro, *Jerusalem Online*, The Contemporary City,
https://doi.org/10.1007/978-981-16-3314-0_1

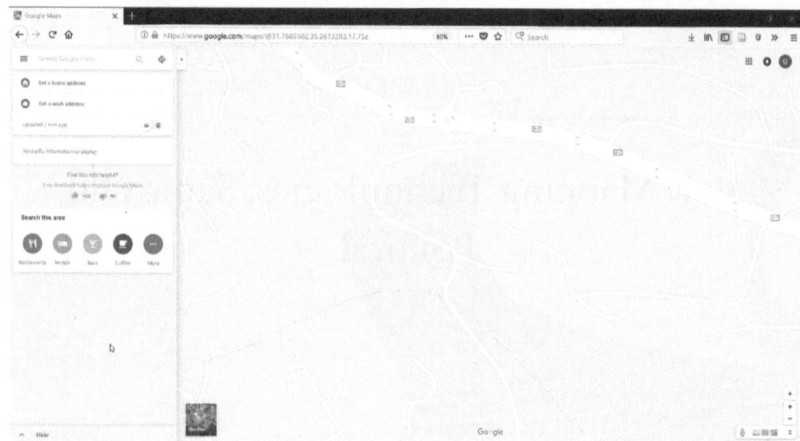

Fig. 1.1 The neighbourhood of Ras al-Amud on Google Maps, as of October 2018 (*Source* Google Maps; Google and the Google logo are registered trademarks of Google LLC, used with permission)

and had recently become Shim'on Ha'tsadik Street, in reference to a nearby Roman tomb traditionally identified as the tomb of Jewish high priest Simon the Just (Murphy-O'Connor 2008). A mismatch between Google's database and the street signs caused countless visitors—myself included—to get lost. On my first day of work as a cartographic assistant for Grassroots Jerusalem (GJ), a colleague had to come and find me in the nearby ultra-orthodox Jewish neighbourhood of Mea Shearim, kindly reassuring me that the incident would not be taken as evidence of poor mapping skills.

Grassroots Jerusalem (GJ) is an organisation striving to amplify the voices of Palestinian Jerusalemites through mapping and other advocacy projects. The office's bookshelves housed well-thumbed copies of classics in critical cartography, from Woods' *The Power of Maps* and Abu-Sitta's monumental *Atlas of Palestine*, to Crampton's introduction to critical cartography. These works masterfully dissect cartography's role in supporting oppressive power systems. Most relevantly for GJ's work, they consider how maps have facilitated, and continue to facilitate, the colonisation of Palestine, first at the hands of the British Mandate and then of the Israeli State. Arab toponyms, depopulated villages, Muslim religious sites, and Palestinian landmarks have been systematically renamed or excluded, as the region has alternatively portrayed as the Christians' Holy Land, the ancestral home of the Jews, or a desert waiting to be settled.

I frequently wondered about the blank spaces on Google Maps and their links to Palestine's erasure from historical maps. For my colleagues at GJ, on the other hand, the reason was self-evident: Google is a profit-driven American company that has nothing to gain from including Palestinians and their perspectives. Since starting this research, I have realised that this way of thinking is very common, and not only among Palestinians: it is taken as a given that maps favour the powerful. Whether right or wrong, this explanation is too vague to be satisfying. The maps we use may well be as power-laden as the ones drawn by British colonial surveyors, but not in the same ways. How exactly does Google profit from excluding Palestinian neighbourhoods? Why do the same omissions also occur in maps by other providers, including those created through collaborative platforms such as OpenStreetMap? Who takes these decisions? How are these political dynamics coded into the algorithms and geodatabases that power online maps?

'THEY HAD THE PLANS, THE TERRITORY, THE MAPS...': CARTOGRAPHY AND THE STRUGGLE OVER ISRAEL/PALESTINE (1923–2000')

Critical cartography has taught us to view maps as at once products and tools of dominant social forces, rather than as faithful representations of an external reality. This way of thinking has been extremely productive, calling into question cartography's claim to objectivity and recognise its role in supporting oppressive power arrangements. This line of inquiry has uncovered the links that tie map-making with territorial expansion, colonialism, nationalism and state-building, and the case of Israel/Palestine[1] stands as a powerful example of these links. A brief overview of the region's recent history as seen through the mapping lens

---

[1] I use the term Israel/Palestine to refer to the areas on which Israel was established in 1949 (sometimes termed 'Israel proper'), the West Bank and the Gaza Strip. As will become clear, one of my arguments is that the boundaries between these areas are blurrier and more dynamic that lines on a map tend to imply. With the phrase Israel/Palestine, I intend to emphasise this fuzziness, rather than present my account as 'balanced' or neutral.

will put the rest of the book into context and clarify what is at stake in today's mappings. With few exceptions (Matar 1999; Quiquivix 2012), historical accounts of the region's cartographic histories tend to start with the arrival of the British in Jerusalem, in 1917; if cartography is defined as a scientific, secular approach to map-making, then this decision is legitimate: until then, most Palestinians lived their lives without 'proper' maps. I, too, will focus on this recent history, but it is worth noticing how conveniently this starting point aligns with colonial narratives that describe precolonial Palestine as backwards to the point of non-existence. The gift of cartography appears as one of several 'complementary noble missions' (Al-Hardan 2008) used to justify colonialism by presenting it as progress. Such narratives, as I will argue throughout the book, periodically reemerge, transformed but clearly recognisable, in discussions of contemporary Israel/Palestine.

As the Ottoman Empire disintegrated, European powers moved to secure their interests in the region. The 1916 Sykes–Picot Agreement defined a French and a British sphere of influence in the Middle East, defying British promises to recognise Arab independence after the war in exchange for military support. The agreement granted Britain control over present-day Southern Israel, the Gaza Strip and the ports of Haifa and Acre. Due to its religious significance, the agreement determined that Palestine would be under an international administration, under British 'supervision'. Between 1917 and 1920, Britain consolidated and expanded its control over Palestine through military actions and, in 1923, it formalised it through an official mandate by the League of Nations, the UN's 'ancestor'. Officially, the mandate's aim was to offer 'administrative advice and assistance' to local communities (League of Nations 1924) and one way to provide such assistance was map-making. There is some debate over the Ottoman Empire's cartographic capacities (see Pursley 2015), but it seems safe to say that the British found themselves lacking the kind of maps necessary to control and administer the newly established Mandate.

The rise of Zionism, the movement supporting first the establishment, and then the development, of a Jewish State in historical Palestine, made these demands more pressing. Modern Zionism had developed starting from the late nineteenth century in Europe, when the notion that nation-states should correspond to ethnically and culturally homogenous communities took hold. What distinguished Zionism from other Jewish nationalist movements was that it successfully 'grafted itself onto

British colonialism' (Balthaser, as quoted in Lazare 2020), using Mandate Palestine to establish itself in the region. This relationship was sanctioned by the 1917 Balfour Declaration, through which the British government expressed its support for Zionism, and its commitment to facilitating 'the establishment in Palestine of a national home for the Jewish people'. In this phase, the Zionist movement focused on promoting mass migration and land purchases in Mandate Palestine, a strategy that intensified the need for a cadastral survey to simplify land transactions, manage disputes, and impose taxation. To fill this gap, the Mandate Survey Department, in collaboration with the Zionist Commission in Palestine, carried out the first comprehensive mapping of Palestine, a project described in great detail by Dov Gavish (2005). Designed according to modern standards and measured through triangulation, the survey maps strived to record the land with scientific objectivity. This also involved systematically recording the names of settlements, landmarks, and land plots, the vast majority of which were, predictably, Arabic. Recognising the danger posed by such maps, a Zionist Naming Committee was tasked with assigning Hebrew names to demonstrate that Jews had historical ties to the land (Benvenisti 2002). Evidently, Zionist leaders were quick to grasp the power of cartography: by naming the landscape, they started to model the nation they hoped to create.

After WWII, a weakened Britain struggled to manage the growing tensions between the Arab majority and the Jewish population, the latter of which was rapidly increasing as Jewish refugees fled Europe following the Holocaust. In 1946, Britain announced its intention to relinquish control over the area; the following year, UN resolution 181 recommended the partition of Palestine into a Jewish and an Arab state, with Jerusalem as a *corpus separatum* under UN control (Fig. 1.2). The resolution provoked a wave of violent clashes between Palestinians and Jews, which eventually degenerated into a war with tellingly different names: 'Independence War' for Israelis, 'al-Nakba' or 'The Catastrophe', for Palestinians. The toll on Palestinians was devastating: of the estimated 1,203,000 Palestinians living in the region in 1946 (UN General Assembly 1947), between 250,000 and 300,000 were expelled from their homes between the publication of the UN partition plan and the start of the war (Finkelstein 2003, p. 62); at least 750,000 were displaced during the war (UN General Assembly 2007, p. A/63/13); over 400 Palestinian villages and towns were destroyed (Pappé 2006); and over 17,000 square kilometres of Palestinian land were expropriated (Badil 2002, p. 3). While

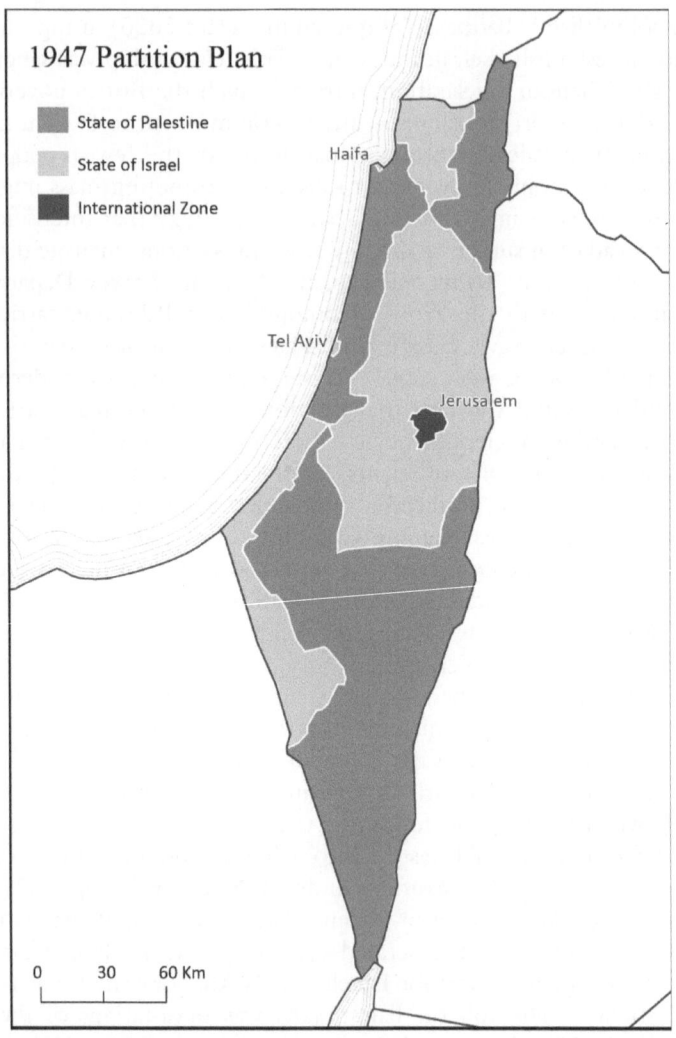

Fig. 1.2 The 1947 UN partition plan (*Source* Map elaborated by the author, based on UN sources)

these numbers and their interpretations are fiercely disputed, scholars (Masalha 2003a; Pappé 2006) have demonstrated that such devastation was not an accident, but rather the result of a systematic plan to make Israel a Jewish state—that is, to 'cleanse it' of its Arab population. The war ended in 1949 with the establishment of the State of Israel, extending over 78% of historical Palestine, while Egypt gained control over the Gaza Strip, and Jordan over what is today the West Bank (Fig. 1.3). The armistice cut Jerusalem in half: West Jerusalem under Israeli rule, East Jerusalem under Jordanian rule (Fig. 1.4). The war ended, but the Nakba did not: the appropriation of land, the displacement and dispossession of people, and the systematic erasure of Palestinian history and identity are ongoing (Khoury 2012; Salamanca et al. 2012).

The Nakba turned the territory into a blank slate, as Meron Benvenisti (2002, pp. 33–37) puts it, on which Israeli authorities could inscribe the names they wished, no longer limited to Jewish-owned private land. Maps were rapidly 'cleansed' of any references to destroyed and depopulated Palestinian villages, and Arab names were Hebraicised or replaced with names taken from the Bible or honouring figures of the Zionist movement. Between 1948 and 1951, the Naming Committee assigned 200 names, as many as it had done since its creation in 1924. The task was treated as an urgent matter, to be completed as soon as possible: the names would become reality only once included in official maps, but new maps could only be printed after the renaming was completed, given that at the time partial corrections or updates would have been costly and impractical. In 1960, the director of the Israeli Survey Department would personally write to the committee chair, requesting that he 'quickly fill in what is missing, especially the names of ruins' and 'increase the pace of the assignment of names so that these will appear as Hebrew maps without defect' (reported in Benvenisti 2002, p. 40). Cartography was being deployed as a 'method of organised forgetting' (Le Guin 2017).

In 1967, after years of tension and clashes, the Israeli Defence Forces (IDF) defeated Egypt, Syria, and Jordan in the Six-Day War, occupying East Jerusalem, the West Bank, the Golan Heights, the Gaza Strip and the Sinai Peninsula (returned to Egypt in 1979) as shown in Fig. 1.5. In a 1980 study, Middle Eastern studies scholar William Harris (1980) estimated that 250,000 people were displaced from the West Bank during the war and in the following months, and 70,000 from the Gaza Strip. Some people fled, some were forcibly removed, others were driven off

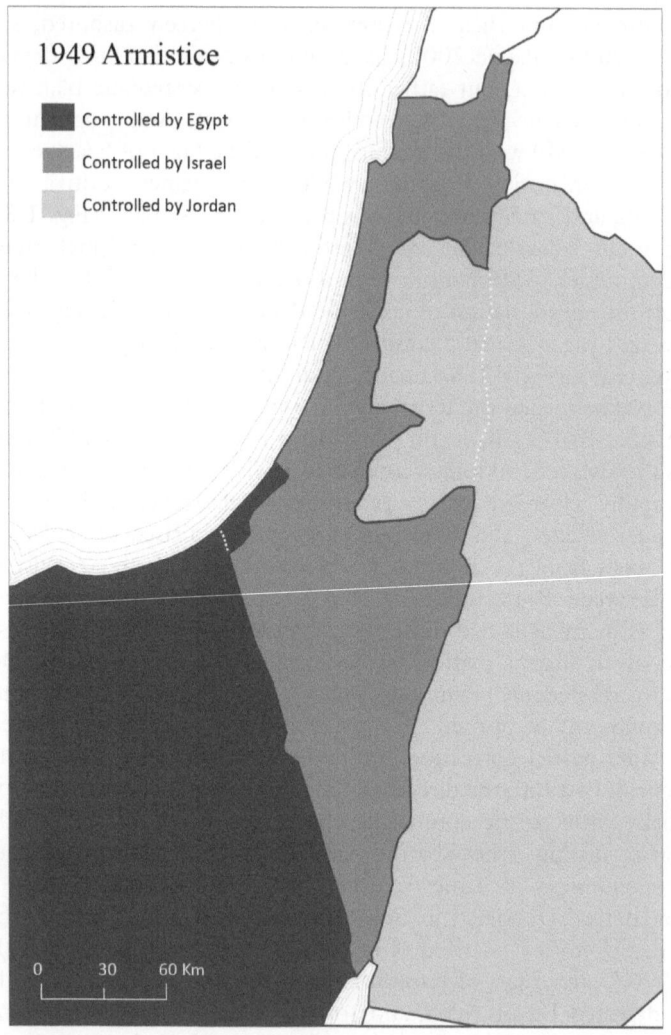

Fig. 1.3 The 1949 armistice lines (*Source* Map elaborated by the author, based on UN sources)

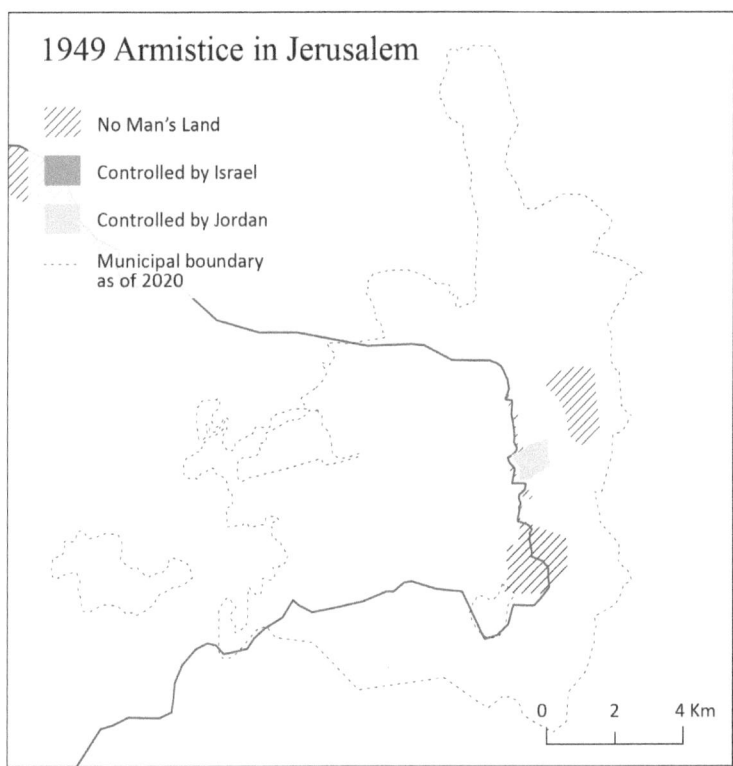

Fig. 1.4 The 1949 armistice lines in Jerusalem (*Source* Map elaborated by the author, based on UN sources)

through demolitions and expropriations, others still were offered monetary incentives to 'migrate' (Masalha 2003b). Nevertheless, contrary to what happened in 1948, most Palestinian inhabitants remained within the newly occupied territories. This posed a challenge to the Israeli state, since annexing the West Bank and Gaza Strip would threaten Israel's Jewish demographic majority. The result was a chasm within Zionist political forces that has persisted until today. Left-wing Zionism envisions the creation of a separate Palestinian state within the 1967 borders, that would exist alongside a Jewish state. Right-wing Zionism, sometimes referred to as Neo-Zionism, considers the West Bank and the Gaza Strip

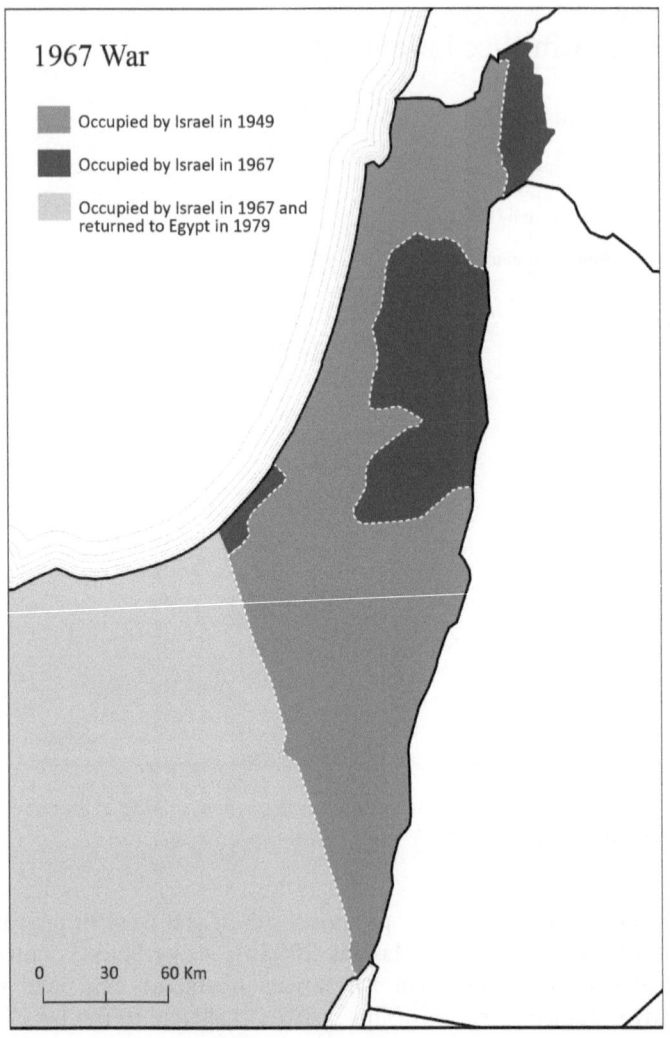

Fig. 1.5 Israel's borders at the end of the 1967 war (*Source* Map elaborated by the author based on UN map 3243)

inalienable parts of the Land of Israel that need to be 'redeemed' (settled) by Jews, and from which Palestinians need to be removed. The Governmental Names Commission, the successor of the naming committee, was called 'to produce toponymical consistency on both sides of the Green Line separating Israel proper [sic] from the West Bank' (Azaryahu and Golan 2001, p. 181) through the Hebraicisation of place names. According to the Commission's estimates, as of 1992, it had issued 7000 Hebrew names. In Jerusalem, this process fully took off starting in 1980, when the Israeli Parliament approved the Jerusalem Basic Law, proclaimed the city 'complete and united', to be the capital of Israel. Meanwhile, using maps to reaffirm Israel's sovereignty on the occupied territories became a government policy, and cartographers were pressured to omit the armistice line from official maps, such as those used in schools (Leuenberger and Schnell 2010, p. 815). This is in contrast with most international maps, which, to this day, reflecting international consensus, depict East Jerusalem, the West Bank and Gaza Strip as occupied territories.

In the 1990s, negotiations between Israel and representatives of the Palestinian Liberation Organisation (PLO) culminated in the Oslo Accords. The agreements, which were supposed to form the basis of a future Palestinian state, envisioned the gradual transfer of control over the West Bank from Israeli authorities to the newly established Palestinian Authority (PA). To this end, the West Bank was divided into three discontinuous jurisdictions: Area A, B and C (Fig. 1.6). Area A comprises major Palestinian towns and is under PA control. Area B includes Palestinian villages and their surrounding lands. It is administered by the PA, while Israel retains control over security. Area C, defined by exclusion as those areas outside Areas A and B, amounts to over 60% of the West Bank's territory and is under full Israeli control. To facilitate a settlement, the issue of Jerusalem was deliberately excluded from the Oslo negotiations.

Many Palestinians felt betrayed by the PLO, since the agreements effectively relinquish claims on the territories occupied by Israel in 1949 and make no provisions for Palestinian refugees and their right to return or receive compensation. As a direct result of the accords, the territory under PA control has been divided into spatially discontinuous territorial units, and Palestinians into an array of political categories: internal refugees, refugees in the diaspora, residents of zone A, B or C and 'Israeli

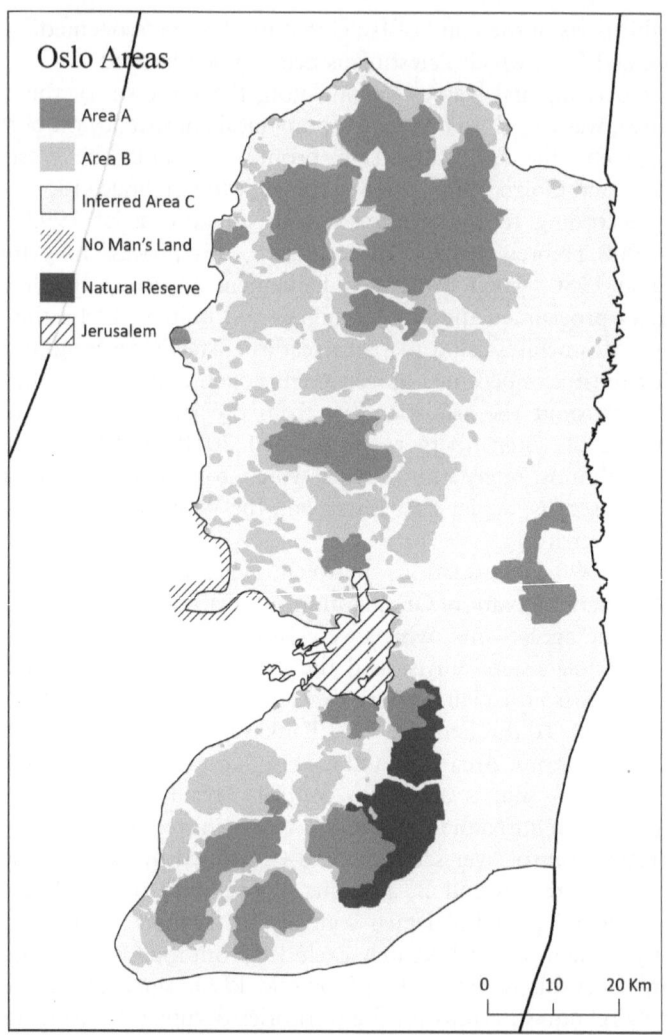

Fig. 1.6 Oslo areas (*Source* Map elaborated by the author based on UN sources)

Arabs' (Hanafi 2009). Characterising Oslo as 'an instrument of Pales-tinian surrender', Edward Said pointed to cartography as a key source of power imbalance in the negotiations:

> The Arab technique has always been to make very large general assertions, and then hope that the concrete details will somehow work out later. (...) They had the plans, the territory, the maps, the settlements, the roads: we have the wish for autonomy and Israeli withdrawal, with no details, and no power to change anything very much. (Said 1995, p. 28)

In the years since, the PA has sought to develop an independent carto-graphic infrastructure, but, paradoxically, has limited access to its own territory, as documented in Jess Bier's recent book, *Mapping Israel, Mapping Palestine* (2017). Israel imposes restriction on aerial photos and satellite imagery of parts of the West Bank, while also preventing Pales-tinian cartographers from surveying them on site. Limited resources and a lack of coordination between the PA and the many non-governmental organisations engaged in mapping compound the problem, leading media scholar Helga Tawil-Souri (2012, p. 57) to conclude that 'in the modern-day balance of "writing" the territory of Israel-Palestine, the outputs weigh heavily on the Israeli side'.

THE GEOWEB: A NEW MAPPING LANGUAGE

Cartography has changed dramatically in the last decade. Paper maps still exist, but only barely, supplanted by mapping applications that are inter-active, multimedia and customisable. Users can query maps for restaurant recommendations, using keywords—or *tags*—to filter by cuisine, price level or atmosphere. Then, with the same application, they can calculate the shortest way there, compare travelling times by car or bus, consult the metro schedule and finally decide to call an Uber. Map-makers can *mash-up* different data sources, overlaying *opensource*, readymade *map tiles* with their own datasets, or with information *crowdsourced* from users, such as photos, travel suggestions or reviews. Meanwhile, many map providers collect information about users—their location, what food they like, how much they are willing to spend, where they live or work, what means of transportation they prefer—and sell it on to advertisers. Back in 2014, Google Maps already boasted one billion monthly users, and that number has certainly grown in the time since. Considering that most people make multiple map searches every day, it represents a truly mind-blowing amount of data! Digital information technologies have brought about so many novelties, that a twentieth-century cartographer would need a specialised dictionary to come to terms with this new language of mapping.

It would be reductive to think of these developments as 'technological', since the emergence of these 'new' maps is inextricably intertwined with economic, social and cultural shifts that are just as profound. To describe these developments, scholars have coined the term *geoweb* (short for 'geospatial web'). A simple definition of the geoweb is 'the use of location as a means of organizing and searching for information on the Internet' (Castree et al. 2013), although, following the example of many others (Elwood and Leszczynski 2011), I use it much more broadly, intending it to refer to new software and datasets as much as to the new mapping actors and practices that accompany them. For example, the transformation of maps into advertising machines was made possible by ever more powerful processors, but also stems from a reconfiguration of the system of cartographic production. Once the monopoly of the state, map-making has turned into a multibillion-dollar industry, involving private data providers and large tech corporations, as well as amateur mappers driven by curiosity, passion or the aspiration to 'free' cartography from state and corporate power. The way people interpret and consume maps has changed, too. As *The Guardian* put it (McMullan 2014), we are by default at the centre of the world, a blue dot from which everything else emanates: not only streets and buildings, but also our own experiences of the city, as recorded in pictures, clips, posts and tweets. In some important way, maps and other geolocated applications become the interface through which we interact with our surroundings. Perfectly capturing why this dependency can be unsettling, urbanist Adam Greenfield (2018, p. 6) observes that digital information technology 'is simultaneously the conduit through which our choices are delivered to us, the mirror by which we see ourselves reflected, and the lens that lets others see us on a level previously unimagined'.

Tech enthusiasts contend that the geoweb has made cartography more accessible, more useful and more accurate. More accessible because maps are now always at the tip of our fingers: a virtually infinite wealth of information available to everyone with an internet connection, at no additional expense, more useful new and ever-evolving map functions: routing, live traffic updates, warnings about accidents and natural hazards, car-sharing support—you name it, there is an app for that. More accurate because increasingly sophisticated software, big data, satellite and 360-degree imagery are seen as the key to finally recording the reality 'as is', or, in the words of a Google executive, putting together a 'digital mirror of the world' (quoted in McMullan 2014), without ideological distortions and political interferences.

Meanwhile, critical scholars (who have long looked at cartography with a certain suspicion) are more diffident than ever, and their concerns are starting to penetrate popular books, news media and online outlets. Software developed by large corporations is designed to serve their interests, they observe (Graham et al. 2013; Leszczynski 2012; Thatcher 2017). This means, firstly, the extraction of profit with little regard for user privacy. As the catchphrase goes, if you are not paying for the product, you *are* the product, and this very much applies to mapping apps. Even more sinister is that, as people grow dependent on digital technologies in every aspect of their lives, they become more and more controllable (Crampton 2015). Through our devices, governments and corporations are able to influence what information we can access, to spy on us and to manipulate our thoughts and behaviours. In this view, recent technological advancements do not make them more neutral, merely more dangerous.

The maps produced by the Israeli and Palestinian survey departments continue to play a crucial role in the region, supporting military actions, security operations, territorial planning and administration. Yet, since the rise of the geoweb, the maps in our phones have acquired as great an impact, becoming the first information source for most people, whether they live in the region, visit it or simply wish to learn more about a faraway place. One of the premises of this book is that critical cartography provides us with a solid foundation for interrogating these new maps, but that it needs an update to meet the distinctive challenges posed by the geoweb. As a recent *Manifesto for Map Studies* put it, 'the world is changing and the way we understand these changes is itself making new worlds', in the face of these changes, we cannot 'deploying rather tired existing ways of imagining the world and simply applying these to interactive, animated and multimediated contexts' (Dodge et al. 2011, p. 240).

To my mind, the first challenge brought about by the geoweb is that it has become much more difficult to hold someone accountable for cartographic content, since authorship is generally split among several people, companies, institutions and, very often, pieces of software. A typical geoweb map combines information from the national mapping agency and several private providers, complements it with data extracted from satellite imagery by a bot, and finally overlays it with content crowd-sourced by thousands of users. Human map-makers design the map interface and write algorithms to handle this tremendous amount of information, but are unlikely to directly decide how a given region, city or street should be represented. Clearly, researchers need new ways to investigate the roles of these different actors in the making of maps, and likely also new ways of conceptualising their agency.

A second challenge is that the geoweb has dramatically accelerated the pace at which geographic knowledge is produced, edited, formatted and circulated. In the past, cartographic surveying and map-drawing were laborious and costly, taking years to complete. Once a map was complete, there was no easy way to change it. Even with the introduction of mapping software in the 1980s, map-making remained a relatively slow process that culminated with a finished map. By contrast, geoweb maps are updated multiple times a day, sometimes in real-time. Users can pan, zoom in and out, change the aesthetics of the map and even select which data layers to display. Platforms like Google Maps also tailor their product to the individual user, for example by selecting places of interest based on previous search queries. The map on your screen is not the same as the one on mine, which is anyway likely to be different by tomorrow. While notions like 'ideology' or 'discourse' evoke relatively coherent knowledge frameworks, legitimised through argument, the messages circulating on the geoweb tend to be fleeting and even contradictory. Maps have become much more unstable, susceptible to change every time they are searched, re-rendered, modified or shared.

As a final challenge, many of the key sources that would help us make sense of the geoweb are difficult to access. Often, the databases and algorithms that power Google Maps and similar commercial mapping applications are copyrighted or protected by trade laws that demand they be kept secret. Even if this were not the case, increasingly complex code would likely make them unintelligible to most people. To make matters worse, the employees of these companies are bound to strict confidentiality agreements and are rarely available for interviews. To gain insight into such opaque systems, researchers need to broaden their inquiries, locating new sources.

Responding to these challenges, this book examines draws on recent work in cartographic theory and feminist technoscience to examine the maps of Jerusalem as processes that acquire their meaning and exert their power through practice. This framework makes it difficult to think of maps as either 'digital mirrors of the world', as Google executives would like us to do, or 'images of power', as proposed by critical cartographers. Furthermore, it illuminates the transformations that mapping technologies undergo as they become embedded in different localities, highlighting the enmeshments of the geoweb and offline realities.

Each of the following chapters focuses on a different mapping provider and a different geoweb feature: distributed authorship on Google Maps, algorithmic navigation on Waze, and crowdsourcing on OpenStreetMap. Building on previous contributions that point to controversies as privileged sites for social inquiry (Latour 2005; Marres 2005), I take as a starting point the Jerusalem-related disputes that surround these maps and use them to explore the political effects of the geoweb on the city. The choice of Jerusalem is largely a personal one, as should already be clear: my experience in the city has shaped my understanding of cartography and inspired this project. In some respects, Jerusalem may seem like an exceptional case, given its disputed geopolitical status and the political tensions that pervade every aspect of urban life. But, it would be a mistake to conclude that its experience is, as Jennifer Robinson puts it (2016), incommensurable. Far from being unique, the ethnic, religious and class conflicts that surround Jerusalem—its resources, its identity, its governance—arise more and more frequently in less polarised cities (Rokem 2016). At once acute and longstanding, such conflicts (and their cartographic manifestations) make Jerusalem a powerful example of how inescapably political maps remain, showing that no amount of artificial intelligence and satellite imagery can represent the world in neutral terms. At the same time, certain parts of my argument are unavoidably specific to Jerusalem and resist generalisation. But I have come to see this limitation as a helpful one. Too often, the digital urbanism scholarship describes the digital or smart city in universal terms: a non-specific city with the same characteristics anywhere in the world. By contrast, here I propose a more situated approach that emphasises the contingency of digital information technologies, recognising that the same feature can do different things in different sites.

BOOK OVERVIEW

The goal of this chapter was to explain how I became interested in maps and their politics, and give readers some background about the role that cartography has played in the history of Palestine, and Jerusalem in particular. The rest of the book is structured into five chapters. Chapter 2, *A Critical Cartography of Sensors and Algorithms*, considers some of the implications of geoweb technologies for cartographic theory and practice. Building on recent theorisations of maps as processes, I propose to view geoweb mappings as entanglements of technological objects and representational practices that blur the distinction between maps and mapped worlds. A critical cartography for the digital age, I suggest, needs to develop new strategies to navigate this complexity. It should approach maps as lively matters that change as they interact with users and non-users, other (digital) media, places and events. However, it should also attend to the durable processes of symbolic and material differentiation enacted through mapping, questioning who benefits and who loses from current arrangements. In addition, I briefly explain the methodologies and sources that underpin my research.

The following three chapters examine three Jerusalem-related cartographic controversies. Chapter 3, *Into Dangerous Territory with Waze*, focuses on the debates surrounding the navigational app Waze in Israel/Palestine and, more specifically, its 'Avoid Dangerous Areas' function, which routes users around areas under formal control of the Palestinian Authority. Against prevalent interpretations that accuse this kind of app of stigmatising racialised, low-income groups and reinforcing socio-spatial segregation, I argue that in Jerusalem Waze has the paradoxical effect of 'putting Palestinians on the map'. By doing so, Waze unwittingly highlights the contradictions inherent to the Israeli state project, which portrays Palestinians as national enemies while simultaneously denying their existence.

Chapter 4, *A Glitch in Google Maps*, looks at the online uproar following the alleged erasure of Palestine from Google Maps in 2016 and uses this as a starting point to discuss the reconfiguration of map authorship engendered by geoweb technologies. The chapter explores how companies, users, algorithms and media platforms contribute to define what appears on Google Maps, and how people interpret this material. In this case, the Google glitch mobilised public attention for a situation that has long been ongoing, namely Palestine's absence from maps of the region—that is, its lack of sovereignty. In a curious twist on the historical relation between cartography and state-building, this 'gap in the map' was taken up on Twitter by Palestinian users in the diaspora, allowing the performance of a national public outside Israel/Palestine territory, and thus showing that maps continue to play an important role in supporting national imaginaries, albeit in distinctively 'postmodern' ways.

Chapter 5, *Naming Jerusalem on OpenStreetMap*, discusses the politics of toponomy in Jerusalem on the crowdsourcing platform Open-StreetMap. While OpenStreetMap is collaborative and ostensibly open to all, its representation of Jerusalem is surprisingly one-sided. The absence of Palestinian names from the map could be interpreted as an objective reflection of reality 'on the ground', or as a sign that Palestinian perspectives continue to be excluded and erased. Instead, drawing on my experience working for GJ and on the analysis of historical sources, I argue that they point to the ongoing confrontations over both the purpose of mapping and the city's identity. Through its mapping standards, OSM intervenes on these confrontations in ways that tend to consolidate the municipality's naming power, but also show its limitations.

Finally, Chapter 6 links together the three case studies, drawing some overall conclusions. The common theme that emerges from the three examples, I suggest, is that Palestinians in Jerusalem are often 'gaps on the map'. While it is certainly possible to read their absence as evidence of political exclusion and marginalisation, I argue that it may be more accurate to interpret it as an instance of 'resistance to being enrolled' (Star 1990) into the Israeli municipality. A more complete and balanced representation of the geoweb will not be particularly helpful in the absence of broader changes to Jerusalem's political and spatial organisation.

References

Al-Hardan, A. (2008). Understanding the Present Through the Past: Between British and Israeli Discourses on Palestine. In R. Lenţin (Ed.), *Thinking Palestine*. London and New York: Zed Books; Distributed in the USA by Palgrave Macmillan.

Azaryahu, M., & Golan, A. (2001). (Re)Naming the Landscape: The Formation of the Hebrew Map of Israel 1949–1960. *Journal of Historical Geography*, 27(2), 178–195. https://doi.org/10.1006/jhge.2001.0297.

Badil. (2002). *Survey of Palestinian Refugees and Internally Displaced Persons.* Bethlehem: Badil. http://www.badil.org/phocadownload/Badil_docs/public ations/Survey-2002.pdf. Accessed 14 August 2020.

Benvenisti, M. (2002). *Sacred Landscape: The Buried History of the Holy Land Since 1948.* Berkeley, CA and London: University of California Press.

Bier, J. (2017). *Mapping Israel, Mapping Palestine: How Occupied Landscapes Shape Scientific Knowledge.* Cambridge, MA: The MIT Press.

Castree, N., Kitchin, R., & Rogers, A. (2013). Geoweb. In *A Dictionary of Human Geography.* Oxford University Press. http://www.oxfordreference. com/view/10.1093/acref/9780199599868.001.0001/acref-978019959 9868-e-717. Accessed 28 May 2018.

Crampton, J. W. (2015). Collect It All: National Security, Big Data and Governance. *GeoJournal, 80*(4), 519–531. https://doi.org/10.1007/s10708-014-9598-y.

Dodge, M., Kitchin, R., & Perkins, C. (2011). Mapping Modes, Methods and Moments: A Manifesto for Map Studies. In *Rethinking Maps: New Frontiers in Cartographic Theory* (pp. 220–243). New York: Routledge.

Elwood, S., & Leszczynski, A. (2011). Privacy, Reconsidered: New Representations, Data Practices, and the Geoweb. *Geoforum, 42*(1), 6–15. https://doi. org/10.1016/j.geoforum.2010.08.003.

Finkelstein, N. G. (2003). *Image and Reality of the Israel-Palestine Conflict.* London: Verso.

Gavish, D. (2005). *The Survey of Palestine Under the British Mandate, 1920–1948.* London: Routledge.

Graham, M., Zook, M., & Boulton, A. (2013). Augmented Reality in Urban Places: Contested Content and the Duplicity of Code. *Transactions of the Institute of British Geographers, 38*(3), 464–479. https://doi.org/10.1111/ j.1475-5661.2012.00539.x.

Greenfield, A. (2018). *Radical Technologies: The Design of Everyday Life* (Paperback ed.). London and New York: Verso.

Hanafi, S. (2009). Spacio-Cide: Colonial Politics, Invisibility and Rezoning in Palestinian Territory. *Contemporary Arab Affairs, 2*(1), 106–121. https:// doi.org/10.1080/17550910802622645.

Harris, W. W. (1980). *Taking Root: Israeli Settlement in the West Bank, the Golan, and Gaza-Sinai, 1967–1980.* Chichester: Research Studies Press.

Khoury, E. (2012). Rethinking the Nakba. *Critical Inquiry, 38*(2), 250–266. https://doi.org/10.1086/662741.

Latour, B. (2005). *Reassembling the Social: An Introduction to Actor-Network-Theory* (First Edition). Oxford University Press.

Lazare, S. (2020, July 13). The Forgotten History of the Jewish, Anti-Zionist Left. *In These Times.* http://inthesetimes.com/article/22659/jewish-anti-zionism-israel-palestine-colonialism-annexation-apartheid. Accessed 12 August 2020.

League of Nations. (1924). Article 22—The Covenant of the League of Nations (Including Amendments adopted to December, 1924). http://avalon.law. yale.edu/20th_century/leagcov.asp#art22. Accessed 20 November 2018.

Le Guin, U. K. (2017). A Non-Euclidean View of California as a Cold Place to Be. In *Dancing at the Edge of the World: Thoughts on Words, Women, Places*. New York: Grove Press.

Leszczynski, A. (2012). Situating the Geoweb in Political Economy. *Progress in Human Geography, 36*(1), 72–89. https://doi.org/10.1177/030913251141 1231.

Leuenberger, C., & Schnell, I. (2010). The Politics of Maps: Constructing National Territories in Israel. *Social Studies of Science, 40*(6), 803–842.

Marres, N. (2005). Issues Spark a Public into Being: A Key but Often Forgotten Point of the Lippmann-Dewey Debate. In B. Latour & P. Weibel (Eds.), *Making Things Public: Atmospheres of Democracy* (pp. 208–217).

Masalha, N. (2003a). *The Politics of Denial: Israel and the Palestinian Refugee Problem/Nur Masalha* (1st ed.). London and Sterling, VA: Pluto Press.

Masalha, N. (2003b). The 1967 Refugee Exodus. In *The Politics of Denial: Israel and the Palestinian Refugee Problem/Nur Masalha* (1st ed., pp. 178–209). London and Sterling, VA: Pluto Press.

Matar, N. (1999). Renaissance Cartography and the Question of Palestine. In I. A. Abu-Lughod, R. Heacock, & K. Nashef (Eds.), *The Landscape of Palestine: Equivocal Poetry*. Birzeit: Birzeit University.

McMullan, T. (2014, December 2). How Digital Maps Are Changing the Way We Understand Our World. *The Guardian*. https://www.theguardian.com/ technology/2014/dec/02/how-digital-maps-changing-the-way-we-unders tand-world. Accessed 25 July 2019.

Murphy-O'Connor, J. (2008). *The Holy Land: An Oxford Archaeological Guide from Earliest Times to 1700* (5th ed., rev. expanded). Oxford and New York: Oxford University Press.

Pappé, I. (2006). *The Ethnic Cleansing of Palestine*. Oxford: Oneworld.

Pursley, S. (2015, June 2). 'Lines Drawn on an Empty Map': Iraq's Borders and the Legend of the Artificial State. *Jadaliyya*. http://www.jadaliyya.com/ pages/index/21759/lines-drawn-on-an-empty-map_iraq%E2%80%99s-bor ders-and-the. Accessed 19 February 2018.

Quiquivix, L. (2012). *The Political Mapping of Palestine*. University of North Carolina at Chapel Hill, Chapel Hill. https://cdr.lib.unc.edu/indexablecon tent/uuid:b079da59-f8ff-4ec5-b1e2-9e335bd059c0.

Robinson, J. (2016). Comparative Urbanism: New Geographies and Cultures of Theorizing the Urban. *International Journal of Urban and Regional Research, 40*(1), 187–199. https://doi.org/10.1111/1468-2427.12273.

Rokem, J. (2016). Learning from Jerusalem. *City*, *20*(3), 407–411. https://doi.org/10.1080/13604813.2016.1166699.

Said, E. W. (1995). Facts, Facts, and More Facts. In *Peace and Its Discontents: Essays on Palestine in the Middle East Peace Process* (1st ed., pp. 26–31). New York: Vintage Books.

Salamanca, O. J., Qato, M., Rabie, K., & Samour, S. (2012). Past Is Present: Settler Colonialism in Palestine. *Settler Colonial Studies*, *2*(1), 1–8. https://doi.org/10.1080/2201473X.2012.10648823.

Star, S. L. (1990). Power, Technology and the Phenomenology of Conventions: On Being Allergic to Onions. *The Sociological Review*, *38*(S1), 26–56. https://doi.org/10.1111/j.1467-954X.1990.tb03347.x.

Tawil-Souri, H. (2012). Mapping Israel–Palestine. *Political Geography*, *31*(1), 57–60. https://doi.org/10.1016/j.polgeo.2011.10.003.

Thatcher, J. (2017). You Are Where You Go, the Commodification of Daily Life Through 'Location'. *Environment and Planning A: Economy and Space*, *49*(12), 2702–2717. https://doi.org/10.1177/0308518X17730580.

UN General Assembly. (1947). *Official Records of the Second Session of the General Assembly* (No. Session 2, Supplement 11). New York: United Nations. https://web.archive.org/web/20120603150222/http://domino.un.org/unispal.nsf/9a798adbf322aff38525617b006d88d7/07175de9fa2d e563852568d3006e10f3?OpenDocument. Accessed 14 August 2020.

UN General Assembly. (2007). *Report of the Commissioner-General of the United Nations Relief and Works Agency for Palestine Refugees in the Near East* (No. Session 63, Supplement 13). New York: United Nations. http://www.unrwa.org/userfiles/20100118141933.pdf. Accessed 14 August 2020.

CHAPTER 2

A Critical Cartography of Sensors and Algorithms

Abstract Dynamic, multidirectional and data-driven, digital mappings blur the distinction between maps and mapped worlds. A critical cartography for the digital age should approach maps as lively matters that change as they interact with users and non-users, other (digital) media, places and events. It should also attend, however, to the durable processes of symbolic and material differentiation enacted through mapping, questioning who benefits and who loses from current arrangements. To this end, I approach maps as matters of care that need to be assembled. Empirically, I take as a starting point the mapping controversies surrounding three major map providers (Waze, Google Maps and OpenStreetMap) and connect them to produce an alternative account of mapping politics in Jerusalem.

Keywords Critical cartography · Processual turn · Geoweb · Matters of care

In 2012, I spent a few days in Jerusalem to visit a friend. Four years had passed since my first visit, and I found a major change in the city. The Jerusalem Light Railway, or JLR, had just opened—the blue seats still pristine, the wide, unscratched windows letting in the views as the train skirted along the edges of the Old City. I remember marvelling at

© The Author(s), under exclusive license to Springer Nature Singapore Pte Ltd. 2021
V. Carraro, *Jerusalem Online*, The Contemporary City, https://doi.org/10.1007/978-981-16-3314-0_2

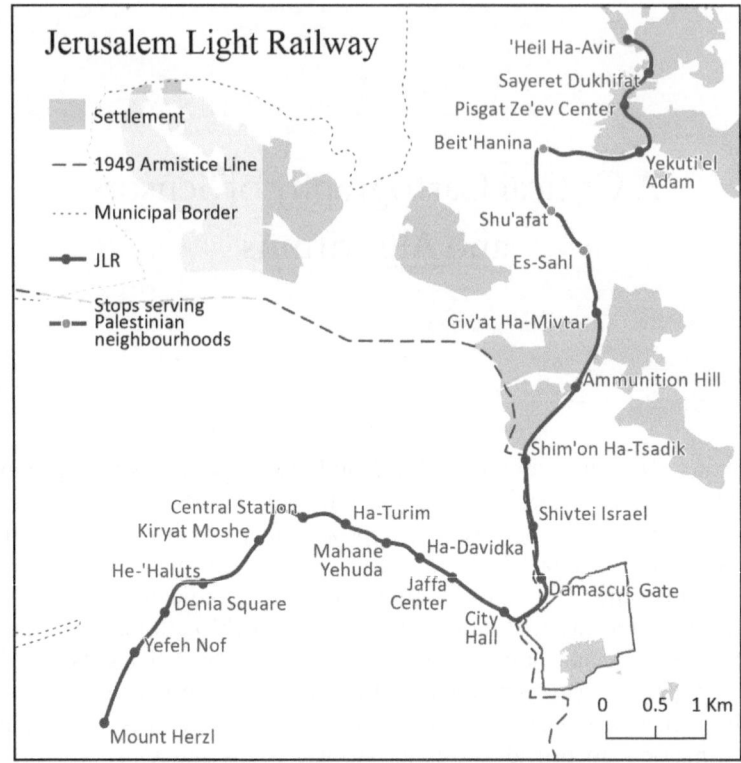

Fig. 2.1   Map of the Jerusalem Light Railway (*Source* Map elaborated by the author using data from OSM [licensed under ODbL] and Peace Now [licensed under CC BY-IGO])

the train's sleekness and feeling reassured by the loudspeaker announcements in English, unaware of the project's sinister politics. Construction works took more than five years, and over a billion, funded through a controversial international consortium that included European and Arab companies. The JLR runs through the city, connecting its western part, under Israeli control since 1948, with the settlements built since 1967 to the south, north and east of the centre, as shown in Fig. 2.1. From the start, these settlements have been built with the aim of consolidating Israeli control over the city, rendering a future partition unfeasible. Military commander and future prime minister Ariel Sharon (quoted in Shaik

1993) said as much responding to a British journalist at Israel's National Press Club in 1973:

> We'll make a pastrami sandwich of them. We'll insert a strip of Jewish settlements in between the Palestinians, and then another strip of Jewish settlements right across the West Bank, so that in 25 years' time, neither the United Nations nor the United States, nobody, will be able to tear it apart.

While only three out of twenty-three stops serve predominantly Palestinian neighbourhoods, the JLR remains the only means of public transport linking Jewish and Palestinian areas, and for this reason, some romanticise it as a metaphor for a difficult but possible coexistence (Badalge 2019; Pfeffer 2012). In fact, coexistence may be a welcome collateral effect, but only so long as it is coexistence under Israeli rule. As Sharon (quoted in Barghouti 2009) made clear thirty years later, reiterating his intention of carrying out the project 'at all costs', the JLR exists 'to strengthen Jerusalem, construct it, expand it and sustain it for eternity as the capital of the Jewish people and the united capital of the State of Israel (…)'. When I returned to Jerusalem again in 2017, the train was decorated with the Israeli flag (shown in Fig. 2.2), in celebration of the fortieth anniversary since the 'reunification' of Jerusalem, making its nationalist significance harder to miss.

JLR passengers today are likely to miss another dimension of the project, namely its cutting-edge combination of artificial intelligence (AI) and real-time data, built by the Israeli start-up, Axilion in partnership with Microsoft. Promotional materials describe the system in flowery newspeak: 'With Axilion cities maximize their current infrastructure via a universal traffic language that bridges multiple hardware vendors for unified communication and smart optimization, while laying the foundation for monetizing congestion' (Cision PRWeb 2019). With a little research and some imagination, one can speculate that traffic lights are equipped with sensors to monitor traffic levels. The data feeds into a central database, which also records the position of trains. Algorithms use this information to coordinate traffic lights, ensuring the trains get green lights while optimising the flow of other vehicles. I remain unsure as to how this 'monetises' congestion, but presumably the term simply refers to the profits made by Axilion and Microsoft through their contracts with

Fig. 2.2 The JLR train in 2017, decorated with the Israeli flag to celebrate the anniversary of the 1967 war and consequent annexation of East Jerusalem (*Source* Photo taken by the author)

municipalities, rather than anything more questionable, such as the calibration of the algorithms to favour certain neighbourhoods or categories of commuters. Overlaying this traffic management infrastructure is a layer of surveillance technology. In 2014, during a wave of unrest (discussed further in Chapter 3), Palestinian protesters repeatedly damaged the JRL stops; in response, the municipality introduced unmanned aerial vehicles (aka drones) and surveillance balloons over the tram stops, streaming high-definition and thermal imagery to a police central aerial unit. As is often the case, in Palestine as elsewhere, the emergency measure has now become routine, coming to form another part of the invisible infrastructure of the JLR (Who Profits 2015, 2017).

Official Israeli narratives around the JLR blend the Axilion-style smart rhetoric with old Zionist imagery. A brochure produced by the Jerusalem municipality describes the JLR as the fulfilment of Theodore Herzl's vision for Jerusalem, boasting 'modern neighbourhoods with electric lines, tree-lined boulevards... a metropolis of the 20th century' (quoted in Barghouti 2009, p. 47). This same vision is also represented in a mural commissioned by the East Jerusalem Development Ltd, a joint government–municipality company, painted in 2001 and still enlivening

Fig. 2.3 Mural of the Jerusalem Light Rail on Jaffa Road (*Source* Photo from Flickr, taken by user Paulina Zet _ Vered Hasharon [licensed under CC BY 2.0])

the corner of Jaffa Road and Strauss Street, on the JLR route. The full-size painting creates a telescopic illusion by depicting a stage nested within the building, with a view of the street as the backdrop, the mural itself visible in the distance. At its centre, is a futuristic train running at full speed. On the stage, an incongruous parade of vintage vehicles, cyclists, a skater jumping off the steps and onto the street, and—most curiously of all, a token camel and donkeys watching the spectacle from the sides (Fig. 2.3). More recent press releases (Jerusalem Municipality 2019), use buzz words like 'smart transportation management', 'urban innovation' and 'green waves' to underscore Jerusalem's advancement, asserting the city's rightful place among 'among the world's leading smart cities'. The JLR fits the broader portrayal of Israel as innovative and dynamic, a nation of 'battlefield entrepreneurs' (Senor and Singer 2011, Chapter 2) who made the desert bloom by sowing 'cyber hubs' (Ackerman 2015). It is a narrative that recycles many familiar Zionist tropes: the resilience of pioneers, the explicit juxtaposition of Israel's 'modernity' and resource-fulness to the Arab idleness and backwardness. It also whitewashes Israel's relative economic success, built in no small measure on the resources and capital produced through the occupation of Palestine (Getzoff 2020). And, crucially, narrative and technology work together: Zionism's colonial vision is sustained through 'a policy of detail' (Said 1992, p. 95), through careful planning, ambitious investments, technological sophistication and fine-grained data.

What fascinates me about this project, and the reason I see it as a good place to begin this chapter, is that it illustrates some of the challenges we face when dealing with contemporary maps. As maps turn into interactive information systems, less like images and more like computer programmes, it becomes difficult to examine them as one would paper maps. Much like physical infrastructures (Salamanca 2014), they sustain certain imaginaries about progress and development, while also materially reconfiguring urban space. Contemporary digital maps like the JLR smart system cannot be said to be the expression of any specific discourse, but this is not to say that they are devoid of meaning or narrative power. Rather, technology (data, sensors, algorithms) and narration (nationalist myths, geopolitical imaginaries, urban branding efforts) are intertwined.

MAPPINGS: FROM BIG NARRATIVE TO BIG DATA?

The JLR smart system may not be the first thing that comes to mind when we hear the word 'map': it is not a finished product, was not authored by a specific person or an institution, is essentially invisible to most people, and does not lend itself to interpretation. Many digital maps, however, share these characteristics: a GPS navigator, the platform through which Amazon matches storage centres and customers, the app that recommends restaurants based on your location. Clearly, the boundary between maps and 'map-like things' has blurred. To describe these systems, David Chandler (2018, Chapter 2) has recently proposed the term *Mapping* (capitalised to distinguish it from the more general practice of making maps), referring to 'an iterative and processual attempt to visualise a particular set of relationships to facilitate problem-solving'. In the JLR example, Mapping traces the relationships between cars, trains and pedestrians and feeds this information to the AI that regulates the traffic light to minimise (and monetise) congestion in a system that constantly updates itself. To emphasise the differences between Mapping and traditional cartography, Chandler draws on recent cartographic theory, particularly work by Rob Kitchin and Martin Dodge. These authors propose that we reconceptualise maps as 'ontogenetic' processes, 'remade every time they are engaged with': every time they are consulted, updated, loaded, panned and searched, interpreted or shared (Kitchin and Dodge 2007, p. 5). This conceptual shift pushes us to see maps as always simultaneously produced and used, serving as visualisations and functional devices, as images and technologies, to the point that 'reality' and 'representation' fold onto one another: map and space are always co-produced. Thus, the map–space pairing merges into one iterative process, always fleeting, contingent and context-dependent.

Kitchin and Dodge are arguably the most influential thinkers in what scholars have recognised as a trend towards non-representational or processual understandings of maps (Caquard 2015; Dodge et al. 2011; Rossetto 2015). I read this trend as a manifestation of the broader ontological turn in social theory (Bryant 2011; DeLanda 2019; Latour 2005a) and as an attempt to explore its implications for cartography. Processual approaches to cartography seek to develop an alternative to the separation between the realms of objectivity and subjectivity, map-maker and map-user, space and map. By doing so, they shift the focus of research from the rules and conditions that shape map knowledge—i.e. epistemology—to the practices and performances through which mappings become—what Kitchin et al. (2013) term 'ontogenetics'.

While the processual turn in cartography seeks to reconceptualise all kinds of maps, Chandler (2018, Chapter 2) notes significant affinities between processual thinking and the principles that underpin Mapping as a novel form of map-making developed in the last decade. In Chandler's interpretation, Mapping emerged as a critique of modernist reason, the notorious 'view from nowhere' so aptly examined by Haraway (1988, 1997). Chandler's argument is that Mapping effectively reverses this logic. While traditional cartography conceives of space in Cartesian terms, as a two-dimensional container filled with discrete objects, Mapping develops an understanding of space as the product of changing relations. While cartography uses statistical categories to order those objects, Mapping relies on sensors, logfiles or online trackers to produce data clusters aggregated 'on the fly'. Where cartography seeks to abstract the world through representation, Mapping is an attempt to index it in an exhaustive manner. There is no longer time for conceptual frameworks and argumentation: information is essentially multidirectional, ephemeral, even incoherent. If cartography is guided by absolute ideals of truthfulness and objectivity, Mapping disavows universal forms of representation and causality, 'bringing knowing closer to reality' (Chandler 2018, Chapter 2, Introduction). As communication accelerates, information replaces discourse as the predominant form of knowing (Lash 2002), Big Data comes to take the place of Big Narrative (Morozov 2013).

Scholars adopting a processual approach to cartography have typically focused on the micro-interactions through which map-making and map-usage unfold: studying the recordings of users consulting a map (Brown and Laurier 2005), interviewing tourist industry workers about their map-mediated encounters with visitors (Del Casino and Hanna 2005), reconstructing how people have engaged with geographical atlases through historical research (Dora 2009), joining a 'mapping party' drifting through Lima, Peru (Gerlach 2015). The premise is that neither the quantitative and experimental methods traditionally deployed by technical cartographers, nor critical cartography's deconstructive analyses can capture 'the unfolding and contextual practices of mapping' (Kitchin et al. 2013, p. 15), and that it is thus time for map studies to broaden their methodological repertoire and, thereby, also their scope. These are certainly useful suggestions for critical cartography,[1] a discipline that has repeatedly been accused of excessive fault-finding, inaccessible language and virtue signalling (Monmonier 2004, 2016). A focus on local practices and change can restore a productive sense of curiosity and possibility: rather than 'explaining away' maps as the product of generic and obscure forces (Latour 2004a, 2005a, pp. 22–23)—'power', 'colonialism', 'structure', 'hegemonic discourse'—researchers are pushed to interrogate how *this* map works for *this* audience at a given time and place.

On the other hand, maps of Jerusalem have always struck me as 'durable' and relatively unchanging. Latour (1987, Chapter 6) famously describes technologies like maps as immutable mobiles, referring to the use of fixed standards and techniques to circulate knowledge (and thus power) across space and time. Certainly, projection, grids, north arrows, conventional symbols and colour associations live on through centuries, rendering online maps recognisable as the offspring of old atlases. Yet, besides these technical elements, I am also thinking of the stories these technical elements help to tell. Anaheed Al-Hardan (2008, p. 236)

---

[1] To be clear, this criticism applies not only to critical cartography, but to many critical branches of the social sciences. Blomley (2006), a self-identified critical geographer, laments his own field's tendency towards pomposity, naivety and sloppiness, while literary and cultural studies embrace the idea of 'postcritique' (Anker and Felski 2017).

suggests that discourses on Palestine are characterised by 'a conspicuous genealogical resilience and continuity', with the same themes repeating themselves in the narratives by European Orientalists, British colonialists, twentieth-century Zionists and, today, the Israeli state. Without necessarily being as coherent and static as Al-Hardan implies, the stories we tell about Palestine and Jerusalem certainly rely on common tropes: the exceptionalism of a holy, ancestral land as warrant for territorial claims, the empty deserts to be tamed by Western pioneers, be it thanks to their agricultural skilfulness or tech entrepreneurialism; and the struggle between two opposing nations aspiring to the same land, like two brothers longing for peace but incapable of reconciling their differences. And, crucially, these stories have not been produced and maintained through sheer imagination but, rather, through 'a slow and arduous process, often through war and always through the use of power' (Pursley 2015, para. 28). Maps have facilitated the process, feeding the imagination and aiding the exercise of power. Digital mapping technologies—crowdsourced photo galleries, smart sensors, AI platforms—draw on these tropes, too. As the JLR example suggests, they often become part of new versions of the same stories or, sometimes, they take them in new directions. To my mind, then, the question is not whether maps have a representational dimension—they do!—or whether it is possible to set this aside in order to focus on practice—it is not!—but, to borrow Timothy Mitchell's piercing formulation, it is a question of 'how one situates the representational in relation to all the other aspects of our complex forms of collective existence and, in particular, to not approach the representational as one half of life' (quoted in Abourahme and Jabary-Salamanca 2016, p. 749).

Processual cartographers see the immutability of maps as an illusion born out of sheer repetition. They tend to avoid or dismiss the term 'object' in relation to maps, associating it with fixity, self-coherency and completeness. To treat the map as an object, they argue, necessarily means to rush to an assumption about its nature, to downplay its potential for change, and, ultimately, to 'move away from our proper engagement with the politics of the map' (Crampton 2011, p. 34; see also Kitchin and Dodge 2007). Maps unfold through creative, affective and habitual practices enacted to solve specific problems, and these practices, even if ephemeral and contextual, give maps 'the semblance' of ontological security (Kitchin et al. 2013, p. 2). Cartographic research needs to focus on these practices, and this means looking 'beyond the map', 'beyond the power of material artefacts and fixed public images' (Dodge and Perkins 2015, p. 38).

I wonder if, in this new-found fascination for the emergent and fleeting, there might be a risk of overlooking the big things that are so durable as to seem timeless. Such tensions—between, on the one hand, fluidity and ontological flatness, and on the other hand, the longevity of historical mechanisms of stratification and differentiation—have been noted before in different but related fields of inquiry, notably geography (Brenner et al. 2011; Kinkaid 2019; Wachsmuth et al. 2011), geopolitics (Sharp 2020), science and technology studies (Haraway 2016; Puig de la Bellacasa 2011) and literary studies (Love 2017). Flat ontologies bring richness and liveliness to critical cartography, directing attention to aspects and openings overlooked by previous scholarship. Yet the insistence on openness and indeterminacy is assumed a priori of any empirical investigation and often overstated. Perhaps more importantly, the celebration of 'on-the-moment-ness' obscures the most important issue, namely that change is exactly what allows power differences to persist (Saldanha 2012, p. 196).

Approaching Maps as Matters of Care

Processual cartographers' treatment of objects stands in stark contrast with work in science and technology studies (Latour 2004b, 2005b; Marres 2005) and geography (Barry 2001; Braun and Whatmore 2010), which argues that politics is, at least in part, a process whereby objects become a focus of disagreement, thereby generating a public. Informed by twentieth-century American pragmatism (Dewey 2012; Lippmann 1993), scholars adopting this materialist understanding of politics view disagreement (as opposed to deliberative consensus, as in the liberal tradition), as the core of democratic politics. This framing aligns with contemporary theories of radical democracy (Mouffe 2005a, b), which posit that, if 'the social is the realm of sedimented practices', political situations are those where the contingency of that order is revealed and questioned (Mouffe 2005b, p. 804). Scholars of material politics, however, are especially interested in the materials, objects and technologies that frequently stand at the centre of political controversies.

New technoscientific objects are often catalysers of political situations because, having just been invented, they are not yet taken for granted. Latour calls these 'hybrid objects' or *matters of concern*. Matters of concern are 'risky attachment, entangled objects', that 'have no clear boundaries, no well-defined essences, no sharp separation between their own hard kernel and their environment' (Latour 2004b, pp. 23–24). They stand in contrast with *matters of fact*, i.e. well-defined, risk-free objects to which we are accustomed. Latour's examples of matters of concern include the ozone layer, the HIV virus (1993), the Columbia shuttle (2005b), global warming, DNA probes and river pollution (2007). To this list, we could add many recent map-related developments: geo-surveillance, autonomous vehicles, the Internet of Things or, indeed, the JLR smart system. What these objects have in common is that they do not conceal the processes through which they are fabricated: we can all see that they have been invented, and are being produced, validated, tweaked and, frequently, fiercely contested. Technoscientific controversies are not always recognised as political: their framing as either political or technical problems is often a crucial aspect of the dispute (Barry 2013, Chapter 1). For Latour, matters of concern offer an opportunity to resist this distinction: because of their hybrid nature, they force us to dispense with the modernist division of power between science—that supposedly deals with how things are—and politics—that deals with 'the representation that human beings make of them' (Latour 2004b, p. 12).

A focus on controversies, rather than maps or mapping platforms, foregrounds the actions through which people engage in mapping: the viewing, clicking and using, the interpreting, arguing and sharing. It explores, on the one hand, how maps affect the public and intervene on existing debates and problems and, on the other, how users react to maps and what they do with them. This framing makes a remarkably good fit for a processual understanding of maps, since it emphasises their dynamism and enactment through practice. Rather than looking 'beyond the power of material artefacts', as advocated by processual cartographers (Dodge and Perkins 2015), this approach encourages us to reimagine objects as lively: not 'arbitrary receptacles of full-fledged society', but actors that can transform, redefine, redeploy or betray what they transport (Latour 1993, Chapter 3). There are differences between maps surveyed with a turning board or satellite imagery, between maps that are engraved or rendered, between map-employing scrolls or touch screens. One can consider these differences while also paying attention to the transformations objects undergo as they become embedded in specific places, in the sort of 'ontogenic', co-productive relation that processual cartographers are so good at capturing. The point is not to settle once and for all what is the map's essence, but to consider what difference mapping technologies make.

The following chapters take as a starting point a series of map-related controversies related to Jerusalem. Depending on the case, they take the shape of public debates, mediatic stunts, disputes on relatively obscure online forums or social media storms. It is not always clear what is at stake, as multiple problems overlap or replace one other in rapid succession. They involve factions of map users, geospatial companies, mapping experts, concerned citizens, journalists, local politicians, even the military. They spark from seemingly trivial matters: a technical glitch, a single map label, unfounded rumours circulating online. They quickly expand, however, attracting public attention and touching upon categorically non-trivial questions about cartography, web technologies and their role in supporting (or not) the political arrangements through which Jerusalem is ruled. Each of these controversies relates to a different mapping provider: Waze, Google Maps and OpenStreetMap. While there are many other geoweb providers, these are among the most used, at least in Jerusalem, according to a small survey I carried out in 2017 (see Appendix). I was also interested in including mapping providers with different characteristics, for example in terms of data sources, typical users, and revenue models. Google Maps needs no introduction: used by over a billion people every month, it is easily the most popular map app worldwide (Panko 2018). Waze is a navigational app, developed in Israel and purchased by Google in 2013. In addition to providing routing and real-time traffic information, Waze integrates a reporting system that allows users to share information about accidents, traffic jams, road closures, construction works, speed traps, potholes and fuel prices, among other things. OSM is the most notable example of a user-led, open-source mapping platform. Its data is used by governmental and international agencies, research institutions and companies such as Uber and Strava. To identify and explore relevant controversies I have used a combination of field research and digital methods, analysing hundreds of news media articles, social media posts and threads on mapping forums. Readers can find a more comprehensive overview of methods and sources in the Appendix.

What makes these episodes significant is not their specific focus, nor the fact that they managed to 'trend' on Twitter for a few hours. Rather, as Barry (2012, p. 330) explains, it is their relations to 'a moving field of other controversies, conflicts and events, including those that have occurred in the past and that might occur in the future'. That is to say, they are not 'pauses' in the struggle for domination, but happen in a mesh of 'force relations' (Meehan et al. 2013) and, in the case of Jerusalem, on an extremely uneven political terrain. Bier's (2017) insightful examination of the asymmetrical power of Israeli and Palestinian cartographers is an obvious and pertinent example here, but the 'power geometries' (Massey 1999) to which I refer are not exhausted by this blunt juxtaposition. Competing visions of their respective national pasts and futures run through both Israeli and Palestinian society, some marginal, some hegemonic, some supported by the State, some by international organisations, some through hefty private donations, some running on a small budget. The web may reconfigure the playing field, but does not level it, as digital infrastructures inform and are informed by geopolitical relations (Aouragh and Chakravartty 2016).

I believe that a study of Mapping in Jerusalem must not only take these power differentials into account, but also challenge them. In her constructive and influential critique, feminist scholar Maria Puig de la Bellacasa (2011) rightly notices that Latour's formulation takes disagreement as the starting point of democratic politics, but stresses consensus building as the only appropriate means to deal with said disagreement. Thus, his emphasis on respectful debate and deliberation quickly turns into an argument to moderate critical standpoints and delegitimise oppositional knowledge, especially when expressed with anger and outrage. These points seem crucial in relation to Jerusalem and Israel/Palestine. Framing the Palestinian question as a disagreement that could be peacefully resolved if only all sides had more respect for one another is deeply problematic, as it disregards the fact that, for some of the parties involved, the current situation is not only perfectly acceptable, but also profitable. In addition, the celebration of dialogue and deliberation as the only forms of democratic politics feeds into the ongoing attack on the Palestinian resistance movement, and in particular, the Boycott, Divestment, Sanctions (BDS) movement, which, as I will discuss in Chapter 5, advocates for a cultural boycott of Israel. As Puig de la Bellacasa (2011, p. 91) puts it, 'we cannot throw out critical standpoints with the bathwater of corrosive critique'.

To reconcile matters of concern with critique, Puig de la Bellacasa draws on the notion of care as elaborated by feminist scholars. *Care* has stronger affective and ethical connotations compared with *concern*. To care for someone or something implies a deeper relation, a sense of responsibility and, frequently, a practical intervention. This does not offer an upper moral or epistemological ground: there is no guarantee that our caring will lead to more accurate knowledge, nor will caring automatically make us more virtuous. It does, however, indicate a commitment to engage with mapping as a process that unfolds 'in strongly stratified technoscientific worlds'. In these settings, it will not do to study matters of care/concern by simply 'tracing issues' or 'following the actors' (Latour 2005a). Rather, to generate care it is necessary to deliberately seek out the marginal, 'counting in participants and issues who have not managed or are not likely to succeed in articulating their concerns, or whose modes of articulation indicate a politics that is "imperceptible" within prevalent ways of understanding' (Puig de la Bellacasa 2011, pp. 94–95).

Approaching these mapping issues as matters of concern/care means recognising that an understanding of maps as unstable and multidirectional does not erase power differentials between the people and institutions involved in their making and remaking. It means 'to take side', but not in a simplistic, two-sided battle predicated on the assumption of 'for' and 'against', 'mappers' and 'mapped', 'overdogs' and 'underdogs' (Richards 1996). Blurring the lines between map-making and map-usage, between subject and object, will not make cartography truly participatory and democratic, as some processual cartographers have suggested (Del Casino and Hanna 2005, p. 51). It may, however, produce different accounts of mapping that will become themselves part of the mapping process, changing the story (Haraway 2016) and thereby reconfiguring those power differentials, if only by a few inches.

References

Abourahme, N., & Jabary-Salamanca, O. (2016). Thinking Against the Sovereignty of the Concept. *City*, *20*(5), 737–754. https://doi.org/10.1080/13604813.2016.1224486.

Ackerman, G. (2015, January 16). Israel Sows Cyber Hub in Desert to Make Beersheba Bloom. *Bloomberg*. https://www.bloomberg.com/news/articles/2015-01-16/israel-sows-cyber-hub-in-desert-to-make-beersheba-bloom-cities. Accessed 2 April 2020.

Al-Hardan, A. (2008). Understanding the Present Through the Past: Between British and Israeli Discourses on Palestine. In R. Lentin (Ed.), *Thinking Palestine*. London and New York: Zed Books; Distributed in the USA by Palgrave Macmillan.

Anker, E. S., & Felski, R. (Eds.). (2017). *Critique and Postcritique*. Durham and London: Duke University Press.

Aouragh, M., & Chakravartty, P. (2016). Infrastructures of Empire: Towards a Critical Geopolitics of Media and Information Studies. *Media, Culture & Society, 38*(4), 559–575. https://doi.org/10.1177/0163443716643007.

Badalge, K. N. (2019, February 14). The Light Rail That Links a Divided City. *CityLab*. https://www.citylab.com/life/2019/02/jerusalem-light-rail-train-israeli-palestinian-divisions/582718/. Accessed 27 April 2020.

Barghouti, O. (2009). Derailing Injustice. Palestinian Civil Resistance to the 'Jerusalem Light Rail'. *Jerusalem Quarterly, 38*, 46–75.

Barry, A. (2001). *Political Machines: Governing a Technological Society* (1. publ.). London: Athlone Press.

Barry, A. (2012). Political Situations: Knowledge Controversies in Transnational Governance. *Critical Policy Studies, 6*(3), 324–336. https://doi.org/10.1080/19460171.2012.699234.

Barry, A. (2013). *Material Politics: Disputes Along the Pipeline*. Hoboken and Chichester, UK: Wiley. http://ebookcentral.proquest.com/lib/cityuhk/detail.action?docID=1420226. Accessed 6 June 2018.

Bier, J. (2017). *Mapping Israel, Mapping Palestine: How Occupied Landscapes Shape Scientific Knowledge*. Cambridge, MA: The MIT Press.

Blomley, N. (2006). Uncritical Critical Geography? *Progress in Human Geography, 30*(1), 87–94. https://doi.org/10.1191/0309132506ph593pr.

Braun, B., & Whatmore, S. (Eds.). (2010). *Political Matter: Technoscience, Democracy, and Public Life*. Minneapolis: University of Minnesota Press.

Brenner, N., Madden, D. J., & Wachsmuth, D. (2011). Assemblage Urbanism and the Challenges of Critical Urban Theory. *City, 15*(2), 225–240.

Brown, B., & Laurier, E. (2005). Maps and Journeys: An Ethno-Methodological Investigation. *Cartographica: The International Journal for Geographic Information and Geovisualization, 40*(3), 17–33. https://doi.org/10.3138/6QPX-0V10-24R0-0621.

Bryant, L. (2011). *The Democracy of Objects*. Open Humanities Press. http://www.oapen.org/search?identifier=444377. Accessed 13 August 2018.

Caquard, S. (2015). Cartography III: A Post-representational Perspective on Cognitive Cartography. *Progress in Human Geography, 39*(2), 225–235. https://doi.org/10.1177/0309132514527039.

Chandler, D. (2018). *Ontopolitics in the Anthropocene: An Introduction to Mapping, Sensing and Hacking*. London and New York, NY: Routledge.

Cision PRWeb. (2019, November 22). City of Jerusalem, Axilion Smart Mobility and Microsoft Azure Are Finalist at the Smart City Expo Award 2019—Mobility Category. *Axilion*. https://axilion.com/new/city-of-jerusalem-axilion-smart-mobility-and-microsoft-azure-are-finalist-at-the-smart-city-expo-award-2019-mobility-category/. Accessed 30 March 2020.

Crampton, J. (2011). Rethinking Maps and Identity: Choropleths, Clines, and Biopolitics. In M. Dodge, R. Kitchin, & C. Perkins (Eds.), *Rethinking Maps: New Frontiers in Cartographic Theory*. New York: Routledge.

DeLanda, M. (2019). *A New Philosophy of Society: Assemblage Theory and Social Complexity*. London: Bloomsbury Publishing.

Del Casino, V., & Hanna, S. (2005). Beyond the 'Binaries': A Methodological Intervention for Interrogating Maps as Representational Practices. *ACME, 4*(1), 34–56.

Dewey, J. (2012). *The Public And Its Problems: An Essay in Political Inquiry* (M. L. Rogers, Ed.). University Park: Penn State Press.

Dodge, M., Kitchin, R., & Perkins, C. (2011). Thinking About Maps. In *Rethinking Maps: New Frontiers in Cartographic Theory* (pp. 1–25). New York: Routledge.

Dodge, M., & Perkins, C. (2015). Reflecting on J.B. Harley's Influence and What He Missed in "Deconstructing the Map". *Cartographica: The International Journal for Geographic Information and Geovisualization, 50*(1), 37–40. https://doi.org/10.3138/carto.50.1.07.

Dora, V. della. (2009). Performative Atlases: Memory, Materiality, and (Co-)Authorship. *Cartographica: The International Journal for Geographic Information and Geovisualization, 44*(4), 240–255. https://doi.org/10.3138/carto.44.4.240.

Gerlach, J. (2015). Editing Worlds: Participatory Mapping and a Minor Geopolitics. *Transactions of the Institute of British Geographers, 40*(2), 273–286. https://doi.org/10.1111/tran.12075.

Getzoff, J. F. (2020). Start-Up Nationalism: The Rationalities of Neoliberal Zionism. *Environment and Planning D: Society and Space*. https://doi.org/10.1177/0263775820911949.

Haraway, D. (1988). Situated Knowledges: The Science Question in Feminism and the Privilege of Partial Perspective. *Feminist Studies, 14*(3), 575–599. https://doi.org/10.2307/3178066.

Haraway, D. (1997). *Modest_WitnessSecond_Millennium. Female-Man©_Meets_OncoMouseTM: Feminism and Technoscience*. New York and London: Routledge.

Haraway, D. (2016). Tentacular Thinking: Anthropocene, Capitalocene, Chthulucene. In *Staying with the Trouble: Making Kin in the Chthulucene*. Durham: Duke University Press.

Jerusalem Municipality. (2019, November 22). Jerusalem Among the World's Leading Smart Cities for Transportation. *Municipality of Jerusalem*. https://www.jerusalem.muni.il/en/newsandarticles/municipality-news/public-transportt-in-jerusalem/. Accessed 31 March 2020.

Kinkaid, E. (2019). Can Assemblage Think Difference? A Feminist Critique of Assemblage Geographies. *Progress in Human Geography.* https://doi.org/10.1177/0309132519836162.

Kitchin, R., & Dodge, M. (2007). Rethinking Maps. *Progress in Human Geography, 31*(3), 331–344.

Kitchin, R., Gleeson, J., & Dodge, M. (2013). Unfolding Mapping Practices: A New Epistemology for Cartography: Unfolding Mapping Practices. *Transactions of the Institute of British Geographers, 38*(3), 480–496. https://doi.org/10.1111/j.1475-5661.2012.00540.x.

Lash, S. (2002). *Critique of Information.* London and Thousand Oaks, CA: Sage.

Latour, B. (1987). *Science in Action: How to Follow Scientists and Engineers Through Society.* Cambridge, MA: Harvard University Press. Accessed 30 October 2015.

Latour, B. (1993). *We Have Never Been Modern.* Cambridge, MA: Harvard University Press.

Latour, B. (2004a). Why Has Critique Run out of Steam? From Matters of Fact to Matters of Concern. *Critical Inquiry, 30*(2), 225–248. https://doi.org/10.1086/421123.

Latour, B. (2004b). *Politics of Nature: How to Bring the Sciences into Democracy* (C. Porter, Trans.). Cambridge, MA: Harvard University Press.

Latour, B. (2005a). *Reassembling the Social: An Introduction to Actor-Network-Theory.* Oxford: Oxford university press.

Latour, B. (2005b). From Realpolitik to Dingpolitik. In B. Latour & P. Weibel (Eds.), *Making Things Public: Atmospheres of Democracy* (pp. 14–44). Cambridge, MA: MIT Press; Karlsruhe, Germany: ZKM/Center for Art and Media in Karlsruhe.

Latour, B. (2007). Turning Around Politics: A Note on Gerard de Vries' Paper. *Social Studies of Science, 37*(5), 811–820.

Lippmann, W. (1993). *The Phantom Public.* New Brunswick, NJ: Transaction Publishers.

Love, H. (2017). The Temptations: Donna Haraway, Feminist Objectivity and the Problem of Critique. In E. S. Anker & R. Felski (Eds.), *Critique and Postcritique.* Durham and London: Duke University Press.

Marres, N. (2005). Issues Spark a Public into Being: A Key but Often Forgotten Point of the Lippmann-Dewey Debate. In B. Latour & P. Weibel (Eds.), *Making Things Public: Atmospheres of Democracy* (pp. 208–217). Cambridge, MA: MIT Press; Karlsruhe, Germany: ZKM/Center for Art and Media in Karlsruhe.

Massey, D. (1999). Imagining Globalization: Power-Geometries of Time-Space. In A. Brah, M. J. Hickman, & M. M. an Ghaill (Eds.), *Global Futures:*

Migration, Environment and Globalization (pp. 27–44). London: Palgrave Macmillan. https://doi.org/10.1057/9780230378537_2.

Meehan, K., Shaw, I. G. R., & Marston, S. A. (2013). Political Geographies of the Object. *Political Geography, 33*, 1–10. https://doi.org/10.1016/j.pol geo.2012.11.002.

Monmonier, M. (2004). *Rhumb Lines and Map Wars: A Social History of the Mercator Projection.* Chicago: University of Chicago Press.

Monmonier, M. (2016). *A Critique of Critical Cartography.* Essay. http://www.markmonmonier.com/attachments/Critique_of_Critical_Cartography_PDF_w_Headnote.pdf. Accessed 15 January 2020.

Morozov, E. (2013, June 24). Big Data and the End of "Why?" *Slate Magazine.* https://slate.com/technology/2013/06/with-big-data-sur veillance-the-government-doesnt-need-to-know-why-anymore.html. Accessed 17 March 2020.

Mouffe, C. (2005a). *On the Political.* London and New York: Routledge.

Mouffe, C. (2005b). Some Reflections on an Agonistic Approach to the Public. In B. Latour & P. Weibel (Eds.), *Making Things Public: Atmospheres of Democracy* (pp. 802–807). Cambridge, MA: MIT Press; Karlsruhe, Germany: ZKM/Center for Art and Media in Karlsruhe.

Panko, R. (2018, July 10). The Popularity of Google Maps: Trends in Navigation Apps in 2018. *The Manifest.* https://themanifest.com/app-development/popularity-google-maps-trends-navigation-apps-2018. Accessed 21 November 2018.

Pfeffer, A. (2012, January 6). Jerusalem's Public Transport System as Metaphor for Israel in 2012. *Haaretz.* https://www.haaretz.com/1.5159889. Accessed 30 March 2020.

Puig de la Bellacasa, M. (2011). Matters of Care in Technoscience: Assembling Neglected Things. *Social Studies of Science, 41*(1), 85–106. https://doi.org/10.1177/0306312710380301.

Pursley, S. (2015, June 2). 'Lines Drawn on an Empty Map': Iraq's Borders and the Legend of the Artificial State. *Jadaliyya.* http://www.jadaliyya.com/pages/index/21759/lines-drawn-on-an-empty-map_iraq%E2%80%99s-bor ders-and-the. Accessed 19 February 2018.

Richards, E. (1996). (Un)Boxing the Monster. *Social Studies of Science, 26*(2), 323–356. https://doi.org/10.1177/030631296026002006.

Rossetto, T. (2015). Semantic Ruminations on 'Post-representational Cartography'. *International Journal of Cartography, 1*(2), 151–167. https://doi.org/10.1080/23729333.2016.1145041.

Said, E. W. (1992). *The Question of Palestine.* New York: Vintage Books.

Salamanca, O. J. (2014). Road 443: Cementing Dispossession, Normalizing Segregation and Disrupting Everyday Life in Palestine. In *Infrastructural Lives* (pp. 128–150). London: Routledge.

Saldanha, A. (2012). Assemblage, Materiality, Race, Capital. *Dialogues in Human Geography*, 2(2), 194–197. https://doi.org/10.1177/2043820612449302.

Senor, D., & Singer, S. (2011). *Start-Up Nation: The Story of Israel's Economic Miracle*. New York: Twelve.

Shaik, M. (1993, November 17). We'll Make a Pastrami Sandwich of Them. *Green Left* (685). https://www.greenleft.org.au/content/well-make-pastrami-sandwich-them. Accessed 13 October 2020.

Sharp, J. (2020). Materials, Forensics and Feminist Geopolitics. *Progress in Human Geography*. https://doi.org/10.1177/0309132520905653.

Wachsmuth, D., Madden, D. J., & Brenner, N. (2011). Between Abstraction and Complexity. *City*, 15(6), 740–750. https://doi.org/10.1080/136 04813.2011.632903.

Who Profits. (2015). *Eye in the Sky: New Aerial Surveillance Systems and the Jerusalem Light Rail* (Flash Report). Who Profits: The Israeli Occupation Industry. https://whoprofits.org/flash-report/eye-in-the-sky-new-aerial-surveillance-systems-and-the-jerusalem-light-rail. Accessed 27 April 2020.

Who Profits. (2017). *Tracking Annexation: The Jerusalem Light Rail and the Israeli Occupation* (Flash Report) (p. 11). Who Profits: The Israeli Occupation Industry. https://whoprofits.org/flash-report/tracking-annexation-the-jerusalem-light-rail-and-the-israeli-occupation. Accessed 27 April 2020.

CHAPTER 3

Into Dangerous Territory with Waze

Abstract Waze is a navigation app that provides routing, as well as information about traffic, road closures and fuel prices, among other things. In Israel/Palestine, it also offers an 'avoid dangerous areas' feature that allows users to avoid Palestinian areas. In this chapter, I examine the debates and controversies generated by this function. While critics from the US have accused similar apps of reproducing spatial segregation based on race and class, I argue that, in Jerusalem, 'social mixing' is ruled out by the political situation and, above all, by a pervasive bordering system made up of differentiated IDs, segregated roads, road closures and checkpoints. In this context, Waze's main effect is to highlight the crucial discrepancy that exists between a spatial and ethnic definition of Israel.

Keywords Waze · Navigation · Borders · Bordering · Jerusalem · West Bank

Established in 1949 to host displaced Palestinians from Jerusalem and nearby towns, the Qalandiya refugee camp sits uncomfortably at the edge of the city. It is located within the municipal boundary of Jerusalem, but cut off from it by the West Bank barrier, as shown in Fig. 3.1. Its population density is estimated at 35,410 people per square kilometre (UNRWA 2015), higher than Hong Kong, Mumbai or Manhattan. To

© The Author(s), under exclusive license to Springer Nature
Singapore Pte Ltd. 2021
V. Carraro, *Jerusalem Online*, The Contemporary City,
https://doi.org/10.1007/978-981-16-3314-0_3

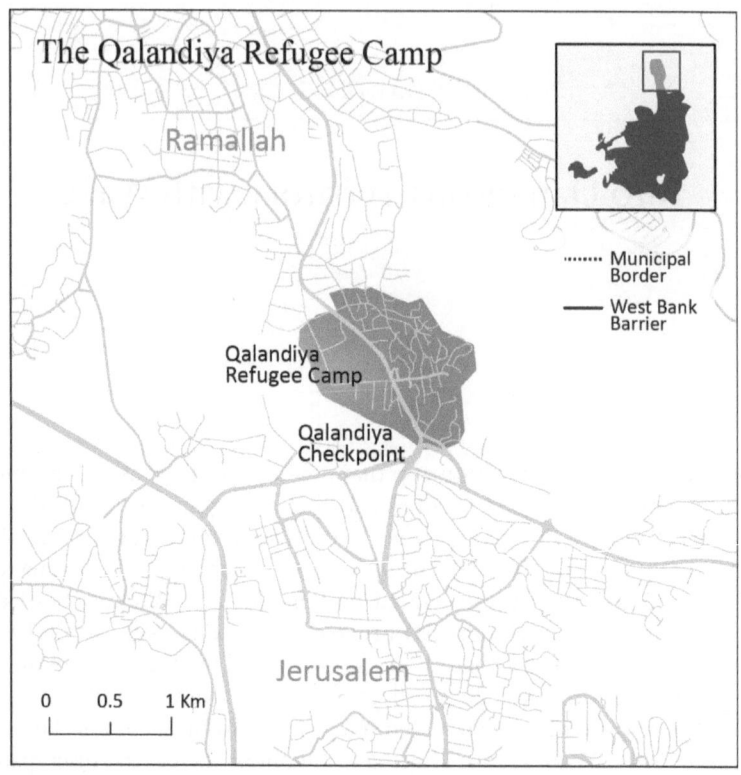

Fig. 3.1 Location of the Qalandiya refugee camp in relation to the Jerusalem municipal boundary and West Bank Barrier (*Source* Map elaborated by the author using data from OSM [licensed under ODbL])

visit the city where they technically live, camp residents must cross the notorious Qalandiya checkpoint, the busiest in the West Bank. Israeli security forces routinely conduct operations within the camp, and just as frequently are met by resistance from the residents. The clashes that ensue, though often deadly (B'Tselem 2013; UNRWA 2015), go largely unnoticed, considered to be, at most, material for local news outlets. In February 2016, however, one of these episodes became an international case. Two Israeli soldiers using the navigation app Waze inadvertently entered the camp, and were promptly surrounded by residents throwing rocks and Molotov cocktails (Katzowitz 2016). The soldiers managed to flee and were soon rescued, unharmed. In the meantime, the army mission sent to their aid engaged in a violent confrontation with the camp residents, killing a Palestinian man and injuring more than a dozen others (Reed 2017). With titles like 'Israeli Soldiers' App Use Leads to Deadly Fight in West Bank Camp' (Beaumont 2016), the news made international headlines. Several reports suggested that, through its faulty security settings, Waze was to blame for the violence; the company responded by revealing that the soldiers had turned off WADA and deviated from the suggested route (Gekker 2016), an account that was later corroborated by an army internal investigation (reported in Finkler 2016). These details notwithstanding, the story lent itself to a 'technological spin' that made the routine of Palestinian casualties newsworthy: did Waze just cause a battle between Israelis and Palestinians?

Danger-Tracking Apps to Keep (White, Middle-class) People Safe

Navigation apps like Waze use algorithms to calculate the shortest route between two points, factoring in variables such as travel distance, road conditions, traffic, one-way streets and so on (for a useful explanation see Byrne 2015). In addition to providing routing and real-time traffic updates, the Waze integrates a reporting system that allows users to share information about accidents, traffic jams, road closures, speed traps and fuel prices, among other things. Adding to this list, Waze 'avoid dangerous areas' function (henceforth WADA) warns users when they are about to enter an unsafe area, offering routes that avoid such places. Although Waze (2016) prides itself on having a large user base in thirty-eight countries, WADA is only available in Israel/Palestine and some parts of Brazil (for a comparison of these local versions see Carraro 2019). The function can be easily turned off, but is activated by default when the app is first installed.

I refer to WADA and other applications designed to warn their users against geographically located dangers as 'danger-tracking' apps. While each danger-tracking app works according to its own, copyrighted algorithms, the basic idea is that their software defines a set of polygons that must be excluded from navigation, as illustrated in Fig. 3.2. Arguably, the first company to conceive of a danger-tracking feature was Microsoft, which registered a Pedestrian Route Production patent in 2012. The

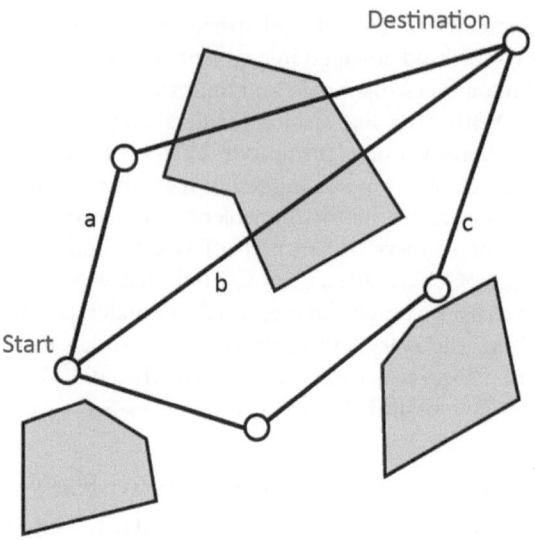

Fig. 3.2 Diagram illustrating the basic working of WADA and similar navigation systems. The app calculates the shortest route from start to destination that avoids crossing the specified polygons. In the example, the only viable option is route c

feature would use information from maps, weather reports, crime statistics and demographics to create pedestrian routes that avoid bad weather, difficult terrain and unsafe areas. Microsoft's system was never made into a functional application, but it had already integrated many of the features that have become common in safety apps, for instance the use of statistical data, the possibility to connect users and calculate routes with a shared destination, or a two-way data flow that gathers data from users as well as serving data to them (Tashev et al. 2012). From the start, people dubbed the system 'avoid-the-ghetto GPS', worried that such an app would penalise poorer and racialised neighbourhoods (Berkow 2012; Milo 2012). Despite these concerns, within just over a year, users could choose from a range of danger-tracking apps, such as GhettoTracker, Safer Route, Road Buddy, SketchFactor and CrimeMapping. Some of these apps cater to drivers rather than pedestrians, draw on user reports rather than statistics, or allow a user to call for help through an S.O.S. button.

The underlying concept, however, is similar: they all use a map interface to guide and protect their users.

Critical geographers tend to see danger-tracking apps as emblematic of the digital city's most disturbing developments (Leszczynski 2016; Thatcher 2013). At the cost of oversimplification, there are two main arguments made against danger-tracking apps. The first is that such apps extend the power that technologies (and the corporations that design and control those technologies) have on our lives. Our daily routes, already influenced by what we see on our phone screens, will be ruled by algorithms that we cannot understand, let alone control. From this perspective, the 'original sin' of danger-tracking apps lies less in their specific functionalities than in their promise to put people's mobility into the hands of private corporations, who will use that power to extract a profit (Thatcher 2013, 2017). The other critique of danger-tracking apps takes issue with the logic that their software uses to divide the city and its inhabitants into categories. Danger-tracking apps are marketed as necessary tools to manage the uncertainty of what is constantly about to happen. Good, law-abiding users are taught to fear the threats that come from 'dangerous others' and protect themselves by using these apps. They contribute to what Amoore (2009) describes as 'war-like architectures' that divide cities into 'us' and 'them', 'here' and 'there', 'safe' and 'risky'.

Nor are the links between safety apps and war confined to imagination: as Amoore points out, we witness an intense exchange of techniques between the military domain and the commercial sphere. For example, algorithms developed to identify patterns of consumption (how likely is it that an individual with these characteristics will purchase product X?) are readapted to guide preemptive security decisions (how likely is that an individual with these characteristics will commit a crime?). Conversely, technologies developed for military use, such as the GPS, move to the private commercial sphere. In Israel, the discourse describing the country as a 'start-up nation' (see Chapter 1) thrives on highlighting connections between the tech sector and the IDF. Special intelligence units are heralded as 'boot camp[s] for startup entrepreneurs', offering veterans 'a fast track into Israeli's high-tech sector' (Orpaz 2015). Tech entrepreneurs are celebrated as symbols of the nation's success: for example, one of Waze's founders was chosen as 'beacon-lighter' for the ceremony on the eve of Israel's 2015 Independence Day (Cashman 2015). And, as of 2020, the IDF is developing a new computer programme to visualise military targets and available striking methods on a map, already nicknamed 'the Waze of Wars' (Gross 2020). Maps have long contributed to create this sort of binary divisions between 'us' and 'them', particularly in colonial contexts (Edney 1997; Ryan 1996). Indeed, following John Agnew (2003, Chapter 1), this is one of the defining characteristics of the global geopolitical imaginary that emerged in modern Europe: the depiction of the world beyond the horizon as 'a source of chaos and danger', the Other against which to define 'ourselves'. In this sense, one could view danger-tracking apps as simply the latest iteration of this process. Yet, their software is based on a novel anticipatory logic, one that works by translating data about past events, e.g. statistics about crime reports, into actionable decisions, e.g. a route. According to digital geography scholar Agnieszka Leszczynski (2016), the problem is that data about past events is culturally and socially informed: shaped, for example, by prevalent definitions of what is a criminal act, by who has historically felt comfortable reporting to the police, and who has historically been perceived as a threat just by virtue of their look. This makes danger-tracking apps inherently conservative, bound to extend into the future today's dynamics of oppression, social stratification and spatial segregation.

Scholarly critiques and popular media commentaries about danger-tracking apps tend to focus, more or less explicitly, on their use in US cities (see Carraro 2019). In that context, words such as 'ghetto' or 'unsafe neighbourhoods' work as euphemisms for poor inner-city areas, generally inhabited by people of colour ('they'). And the branding of some of danger-tracking apps plays on this characterisation, using stock photos to depict users as white, middle-class, white-toothed, traditional families ('us'). Beyond the looks, apps relying on user-contributed data are deemed particularly problematic because of their tendency to reflect the racial bias of their users. Voicing this concern, Joe Silver (2013) of the *American Civil Liberties Union* notes:

> Applications like Waze rely on users and social networking connections to share travel information and suggest particular routes. It's easy to imagine very subtle judgments among loosely connected social groups having a large influence on where drivers and pedestrians are directed in their travels.

The concerns formulated by academics and civil society representatives resonate with, and influence, the perspectives of media: of the eighteen

English-language articles I examined in writing this chapter, fifteen are strongly against danger-tracking apps, with only one presenting some arguments in favour. With titles such as 'Is Your Turn-By-Turn Navigation Application Racist?' (Silver 2013), 'Enough Already with the Avoid-The-Ghetto Apps' (Badger 2013), 'GhettoTracker Is the Worst Site on the Internet' (Holmes 2013) and 'People Just Won't Give Up on Awful Neighborhood Apps' (Capps 2014), it is not hard to gauge the general sense of this criticism.

Such negative press coverage may well be one of the reasons Waze has so far turned down user requests to activate WADA in the US. As a company representative was quoted saying in the Waze forum:

> I know this is an important subject. But the fact is we can't designate an area as dangerous just because we decided to, or even because community members say it is. In Israel some areas are restricted by law, and we follow that law. In other countries situations like these can't be solved by us.[1]

As I am about to illustrate, the public reaction elicited by WADA in Israel/Palestine was very different. The question, then, is whether these critiques can be transposed directly onto Jerusalem. Do people everywhere share the same understanding of urban danger? Do they use danger-tracking apps in similar ways? Do these apps always work as 'future-ing' devices, and if so, do the futures they produce have the same characteristics everywhere?

WHEN 'THEY' ARE 'HERE': NAVIGATING JERUSALEM'S SOFT BORDERS

I was unable to establish WADA's exact launch date, but the feature is mentioned by Israeli users in the Waze forum as early as 2012, the same year Microsoft registered its patent. Information from those early years is scarce, since the feature did not generate much debate, being widely considered a sensible and necessary function in the Israeli context. Contrary to most danger-tracking apps, the Israeli version of WADA defines dangerous areas not by means of statistical data or user reports, but through the geopolitical lines drawn by the Oslo Accords (discussed

---

[1] As reported by user AndyPoms in the 'Waze App Feature Request' Forum, in the thread 'Identificar áreas de risco de assalto' (August 2015).

in the Chapter 1). Initially, Waze's dangerous areas comprised Area A and Area B; arguably, this is also the extent of the Israeli state as experienced by most Jewish Israelis on an everyday basis. For example, virtually no Israelis live in these areas, while approximately 400,000 live in Area C, and 200,000 in East Jerusalem (Peace Now 2016). Israeli citizens are legally forbidden from going into Area A and discouraged from entering Area B, even if these prohibitions are not strictly implemented. Lastly, on a practical, driving related note, Israeli car insurance does not cover travel in Areas A or B.

Waze users have been called WADA into question starting from 2014, when several Israeli users experienced physical attacks or threats while driving through Palestinian areas, especially neighbourhoods in the Eastern part of Jerusalem (e.g. Dvir 2014; Schechter 2014). These incidents happened against the backdrop of a wave of unrest involving demonstrations, rock throwing and, in some instances, stabbings. Worries about the safety of Israeli drivers generated debates in the Waze user forum: some people wished the app would extend its definition of dangerous areas, while others argued that to do so would amount to conceding to Palestinian demands. Area C is 'liberated land', argued one person—a part of Israel just as the Arad or the Negev. Why should Waze exclude it from its navigation system? WADA should cover any route that poses a danger to Israeli drivers, including those parts of Area C targeted by stone throwers, insisted a more cautious driver. Following that logic, countered another, Tel Aviv would also count as a dangerous area: after all, 'most terrorist attacks occurs outside the Palestinian territories since most of the Israelis aren't there'.[2]

Likely responding to these discussions, in 2015 the company initiated a collaboration with the Israeli police and security forces, with the aim of generating dynamic polygons that covered 'hot' Palestinian areas, based on up-to-date security reports. *Yedioth Ahronoth*, one of Israel's leading newspapers, questioned this decision, suggesting that, by doing so, Waze was misrepresenting parts of Jerusalem as controlled by the PA (Yanovsky

---

[2] The views reported here were voiced in several Waze forum threads between 2014 and 2016, under the titles ' ??השטחים דרך לא לסוע אפשר איך' ('How come it is not possible to travel through the territories ??'), ' שטח בין הפרדה B ו-A' ('Separation between the area B and A') and ' מסוכן ככביש 60 כביש הגדרת לגבי צ"מגל פניה' ('An appeal on the Army Radio to set Route 60 as a dangerous area').

2015). The publication also elicited comments from prominent political figures, starting with Jerusalem mayor Nir Barkat, who had recently gained public support by physically intervening to block the politically motivated stabbing of an Israeli civilian (Chandler 2015). Barkat declared Waze's designation to be 'factually incorrect and unacceptable' and urged the company 'to change it and not to turn an app into a political tool'.[3] Aryeh King, member of the Jerusalem City Council and director of the Israel Land Fund, an organisation dedicated to increasing Jewish land ownership beyond the 1949 armistice line, was quoted in the same article as saying:

> Is it conceivable that the police would tell Jews where and where not to go? We are talking about defining areas of Jerusalem as Area A (Palestinian Authority territory). I expect the minister of public security to order the police to set guidelines so that Waze stops, in effect, dividing Jerusalem.

When, a few months later, Waze was implicated in the clashes at the Qalandiya refugee camp, the Israeli left blamed right-wing nationalists for putting pressure on Waze to restrict its dangerous areas. Although users had no way to know the precise boundaries of the polygons, they disagreed on what criteria should be used to define them. For example, some wondered why WADA did not cover the camp (in my understanding, it did),[4] while others insisted that, since the camp is part of Jerusalem, it is the police's responsibility to ensure safety.[5] On Facebook, in a post that gathered thousands of likes, Israeli journalist and blogger Haim Har-Zahav (2016) voiced this sentiment writing: '[…] we showed everyone what Israeli pride is in the app, and we insisted that we would not give up a single inch on the virtual map - it was worth it! Wasn't it?'[6]

Middle East Politics scholar Michael Dumper (2014) characterises Jerusalem as a 'many-bordered city', making a distinction between hard

---

[3] The article, originally published in Hebrew on 08 August 2015, was republished on the Yedioth Ahronoth's English-language website (www.ynetnews.com) on 28 August. All quotes are taken from the English version.

[4] Here, I am relying on the reports of journalists who investigated the issue at the time, since there is no way for me to access previous version of the Waze map.

[5] See for example thread 'Why does Waze let people navigate into refugee camps?' (הפליטים מחנות בתוך לנווט נותן וויז למה ?) on the Israeli user forum.

[6] Automatic translation by Google, with minor edits in the interest of readability.

borders, such as walls, armistice agreements and checkpoints, and soft borders, defined by functional divisions, such as residential or educational segregation. Dumper's point here is that there are frequent mismatches between hard and soft boundaries, so that areas enclosed by the same hard boundaries are subject to different degrees of Israeli control. These remarks provide a useful frame for thinking about the mismatch between the hard boundaries drawn by WADA and the soft boundaries of danger, as they are experienced by Israeli drivers. The two Jerusalem neighbourhoods of Ras al-Amud and Shu'fat (shown in Fig. 3.3) lend themselves as examples, not least because, having lived for six months in each neighbourhood, I am relatively familiar with them.

Ras al-Amud lies at the edge of Jerusalem, delimited to the East by the West Bank Barrier. The neighbourhood is crossed by Jericho Road, the only street to be lit at night. Once the main connection to the Palestinian town of Jericho, the road is now a dead end, blocked by the barrier. Ras Al-Amud is relatively near the city centre—only about a kilometre from the Old City as the crow flies—but it is badly connected by public transport, served only by informal car shares and the minivans run by one of the Arab companies operating in the Eastern part of Jerusalem. These minivans have no fixed timetable, departing when they are full, and stop service in the early evening, usually around 8 pm. There are Israeli buses connecting the nearby settlement of Ma'ale HaZeitim, which have a more regular schedule. The two networks are effectively segregated: few Palestinians use Israeli buses, and virtually no Jewish Israelis use the Palestinian minivans. Jewish taxi drivers, in my experience, categorically refuse to drive into the area, and will instead suggest that their customers change to an Arab taxi at Damascus Gate, where the 1949 armistice line runs. In fact, I never saw *any* Israelis at all in Ras al-Amud, except for soldiers in the armoured vehicle frequently stationed by the main roundabout.

Shu'fat, is located North of the Old City. Surrounded by rapidly developing settlements, it has undergone many infrastructure upgrades in recent years. During my stay, for example, the street on which I lived was paved, provided with house numbers, lined with trees and hooked up to a newly built high-speed Internet network. The most significant upgrade, however, took place in 2011, with the opening of the JLR (discussed in greater detail in Chapter 2). Trains run every few minutes from dawn until midnight and are used by Jews and Palestinians alike. Commuters transit through Shu'fat every day, by both car and train. In fact, it is not unheard of for Jewish Israelis to buy groceries at local Palestinian shops. This is not

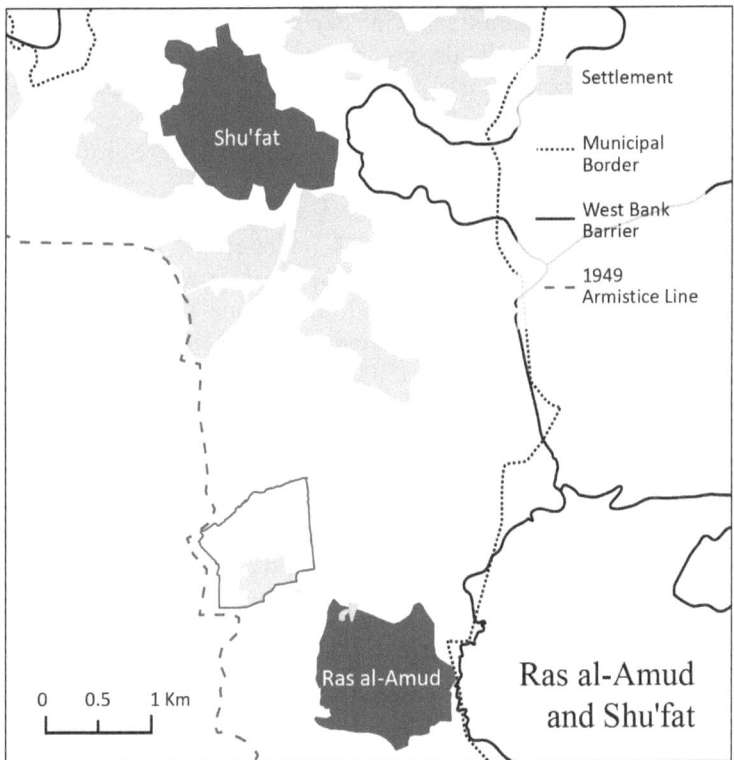

Fig. 3.3 Ras Al-Amud and Shu'fat. The map shows the position of these neighbourhoods in relation to Jerusalem's 'hard' boundaries: the 1949 armistice line, the West Bank barrier and the municipal boundary (*Source* Map elaborated by the author using data from OSM [licensed under ODbL] and Peace Now [licensed under CC BY-IGO])

to say that Shu'fat offers a friendly space for Jewish Israelis. Indeed, their sense of safety is very precarious: tellingly, a Palestinian friend confessed to staring threateningly at Israeli drivers waiting at traffic lights to make them uncomfortable. She reported high rates of success.

In juxtaposing these two descriptions, my intention is to give a sense for the different experience these areas present to their visitors. Although they are both within the Jerusalem Municipality, east of the armistice line but on the Israeli side of the West Bank Barrier, they 'feel' very different.

The soft boundaries delineated by integrated infrastructures and the daily presence of Jewish Israelis do not necessarily match the hard lines of geopolitical boundaries, however, defined. They are fuzzy and unstable, always susceptible to being shifted by an episode of violence, or even the threatening stare of a passerby. It is no wonder that Waze struggles with translating them into polygons on a map.

At first sight, then, the way WADA works in Jerusalem seems to both fit into and reinforce the 'war-like architecture' problematised by critics of danger-tracking apps: a software that entrenches these differences, dividing the city into 'us' and 'them', and then mapping those identities onto space, matching people to specific coordinates: 'we' are 'here', 'they' are 'there' (Amoore 2009). In Jerusalem, however, there is a crucial discrepancy between a spatial definition of danger ('there') and one based on ethnic identity ('they'), or, to flip the perspective, between a spatial definition of Israeli territory ('here') and a definition based on Jewishness ('us'). The tension between these competing definitions lies at the heart of the Zionist colonial project: annexing Palestinian land while simultaneously excluding the Palestinian population.

This project, at once expansionary and exclusionary, has been pursued through a complex system of non-linear borders made up of checkpoints, ID documents and travel permits. It is a system that, as Tawil-Souri (2012, p. 153) observes, 'produce[s] distinct people and bind[s] them to specific territories (such as the Palestinians), while allowing others (Jewish-Israelis) to "trespass" over those same boundaries'. The fuzziness of these boundaries creates the need for an app that alerts Israeli drivers when they are about to cross the 'enemy lines', but also makes it difficult to record dangerous areas as discrete polygons. What is more, by locating Palestinians on a map, the app must acknowledge their presence in places that many users consider non-negotiably Israeli, most notably the heart of Jerusalem. Right-wing Zionists can object to these polygons' discursive implications, but the app stops being useful if it refuses to acknowledge the Palestinian presence on ideological grounds. By recording the borders of the 'Palestinian danger' on a map, WADA draws attention to the fact that 'they' are in fact 'here' (see also Veracini 2013). In this sense, Waze resorts to one of the most classic tools of social control in the cartographer's repertoire. It categorises, measures and records. That is, it forces the messiness of reality into a neat grid, thereby producing spaces and populations that are easier to know, subdue and manage (e.g. Bauman 1998; Crampton 2010; Elden 2007). And yet, this simplification can

sometimes have unexpected effects, such as that of highlighting the inconsistency of a state system that holds Palestinians to be the nation's natural
enemy while also denying their existence.

Palestinian Lives in Danger

Just as Waze's definition of danger in the US starts from the unspoken
premise that the app users are white and middle class (working class
people of colour can hardly avoid the neighbourhoods in which they
live),[7] in Jerusalem Waze assumes its users to be Jewish Israelis.
Challenging this assumption, a *Vice* article (Stuart 2016) published a
few months after the confrontations at the Qalandiya refugee camp
denounced Waze's double standard when it comes to identifying
dangerous area, and proposed that Palestinians, too, should have the
option to avoid dangerous areas while driving. *Vice* conceded that it
would be difficult to create an algorithm able to navigate the West Bank's
'tangled mindfuck of differing jurisdictions, military bases, checkpoints,
barriers, bypass roads and more', but concluded, 'no one is expecting
Waze to be perfect. Being fair is another story'.

Though well-meaning, the suggestion comes across as absurd because
Israeli–Palestinian relations are not symmetrical. Palestinian drivers are
much more likely to experience danger stemming from the well-planned,
pervasive presence of the Israeli security apparatus than they are from
unforeseen encounters with violent Israeli civilians (although those
happen too, see Berger 2018; The New Arab 2018). As the essay
by Tawil-Souri cited above puts it, ID cards generate borders that
'"embrace" Palestinians everywhere and anywhere' (Tawil-Souri 2012,
p. 161), limiting their movements but allowing Jewish Israelis to drive
through. These borders are operationalised through an extensive network
of checkpoints, gates and earthmounds: over 700 according to a recent
survey (OCHA 2018). To be clear, most of these closures are limiting
movement *within* the West Bank, i.e. within areas nominally controlled
by the Palestinian Authority. The same survey also estimates that 400 km

---

[7] Likely, one could make the same arguments about danger-tracking apps in many cities
of the Global North, where racism and spatial segregation can be just as blatant. However,
in the absence of studies focused on these other contexts, I prefer to restrict my focus to
the US.

of roads within the West Bank are 'settlers-only', meaning that Palestinians are banned or restricted from using them. This system of borders, segregated roads and closures is not an unfortunate side effect of Israel's cumbersome but necessary security apparatus. Rather, it is one of the technologies through which settler colonialism works. Infrastructures, such as roads, electricity and telecommunications networks, play a crucial role in the construction of national territories. As geographer Omar Jabari Salamanca (2016) observes, the West Bank infrastructures 'territorialise' the settler community while 'deterritorialising' natives. Danger, intended as the threat of state violence, follows Palestinians wherever they go, including into their own neighbourhoods and homes, as it happens all too often for residents of the Qalandiya refugee camp.

When it became clear that the two Israeli soldiers had entered the Qalandiya refugee camp and that their lives were at risk, Israeli security forces raided the camp with hundreds of soldiers, helicopters, bulldozers and a tank (AFP 2016). Properties were burnt and demolished, people were harmed and, most tragically, 22-year-old Iyad Omar Sajadiyya was shot dead. According to Arab sources, he was one of twelve camp residents to have been killed within a year (Hassan 2016). In the news coverage of the incident, however, these are at most secondary details, because the focus is on Waze: were the soldiers really using the app? Is that something soldiers should be doing? Was Waze to blame?

Discussions about whether the fears of drivers are justified, or whether technology can answer them, tend to monopolise attention. Meanwhile, the lives of Palestinians in Israel (and Black lives in the US)[8] are under far greater threat, not by the random attacks of robbers or stone-throwing protesters, but by the systematic violence perpetrated by the state through police and military forces. To critique danger-tracking apps on the premise that they allow users and contributors to give in to their own biases and racist assumptions minimises the problem. Clearly, living in the Qalandiya refugee camp is a major danger, one that cannot be mitigated by using an app. This is an example of a 'neglected issue' that I wish to highlight through my research. The study of map controversies and their unfolding through user forums and news articles is as much the analysis of what is being said as of what is left out.

---

[8] On racialised policing practices in Israel and the US, and on Black-Palestinian solidarity, see the contributions by Noura Erakat (2020) and Robin Kelley (2019).

* * *

The Israeli version of WADA designates dangerous areas not through algorithms, statistics and user reports, but through good old-fashioned geopolitical lines and military decisions. Its software does not attempt to anticipate dangerous subjects based on data, but simply assumes that Palestinians represent a threat by virtue of being Palestinians, an assumption shared by most of the Israeli public. The problem is that the soft boundaries delineated by integrated infrastructure and a sense of safety for Jewish Israelis do not always match the relatively hard boundaries of the Oslo Agreements. Thus, by recording dangerous areas in the form of polygons, WADA puts on the map the many Palestinians who live in areas that many consider to be part of Israel, such as Area C and, to an even greater extent, East Jerusalem. These discrepancies were highlighted in the aftermaths of the incident at the Qalandiya refugee camp, causing a rift between those who viewed Waze dangerous areas as symbolically legitimising Palestinian territorial claims, and those who believed the safety of Israeli drivers should come first.

This account recast WADA as a technology that unwittingly underscores the fundamental inconsistency of Zionist narratives that portray Palestinians as national enemies while simultaneously denying their existence. By doing so, it pushes against prevalent analyses that question these technologies for reinforcing segregation by minimising the chance of spontaneous encounters (Amoore 2009; Leszczynski 2016; Thatcher 2013). In Israel/Palestine, I argue, this critique loses much of its persuasive power because an extensive bordering apparatus, made up of different types of IDs, barriers, checkpoints and segregated roads, already separates Jewish Israelis and Palestinians, and especially the Palestinians who do not hold an Israeli passport. In the present circumstances, spontaneous encounters between Israelis and Palestinians are neither likely nor desirable.

People from all over the world use the Waze app, but the differences I highlighted to underscore the importance of grounding the study of digital technologies in particular sites, considering the importance of local settings in shaping how these technologies work. WADA does not embody any single, universal logic or mode of governance, but is defined by the associations of different actors: Waze as a company, the technical features of locational apps, users voicing their demands and expectations, politicians performing their commitment to Israel's national project, news and social media, and the broader public. Place-specific discourses around race and security also inform the app, as do the 'material' geographies of territorial agreements, checkpoints, military interventions and policing.

One could see this insistence on local specificities as an academic indulgence, perhaps accurate but politically unhelpful. Granted, some of the themes outlined by the critics of danger-tracking apps remain central to my account, notably the use of a rhetoric based on fear, and the othering of racialised neighbourhoods and their residents. And if Black and Palestinian lives are systematically devalued while being constituted as a threat, and WADA plays into these imaginaries, why dwell on the differences? However, I am convinced that it is possible to critique WADA, in its different iterations, without overlooking the specific circumstances that inform its working in Jerusalem and other places. To articulate such a critique, one does not need a simple narrative that repeats itself everywhere. Indeed, a nuanced understanding of WADA, one that includes multiple accounts that build upon each other, arguably provides a better base for critique. Here, my argument intersects with recent debates between scholars of smart urbanism concerning the current trend to view the smart city as 'a kind of universal, rational and depoliticised project' (Farías and Widmer 2018; Shelton et al. 2015, p. 14). These forms of 'theoretical determinism' (Pow 2015) render scholarly descriptions of technological urban futures more and more similar to the synopsis of films like Minority Report or Blade Runner. Surveillance technologies, we learn, have surpassed George Orwell's bleakest predictions (Haggerty and Ericson 2000). The elites fill cities with smart technologies—devious devices that entrap the rest of us in 'a seamless web of surveillance and power' (Sadowski and Pasquale 2015, p. 1). Meanwhile, the digital has brought about a 'new dark age', leaving us drowning in an overload of information, as our lives are increasingly governed by unintelligible code (Bridle 2018). Danger-tracking apps can seem like tasters of these troublesome worlds to come.

A plurality of accounts helps to remind us that the future has not yet been written. Which technologies take hold, and how they develop, also depends on how people collectively react to them. Admittedly, the process is far from democratic: it is not software development from the ground up, but the nudging of complex systems—informed by market demands, political agendas and uneven relations between socio-economic groups—in one direction or the other. Nor is there any guarantee that the public will 'push' in a progressive direction. Nevertheless, it is heartening to know that public engagement *can* make a difference. As I have mentioned at the beginning of this chapter, the backlash experienced by early 'ghetto-tracking' apps in the US seems to have dissuaded many tech companies, Waze included, from further pursuing the idea, at least for the time being. As of October 2020, Microsoft has not implemented the 'avoid-the-ghetto' function. The GhettoTracker app was first rebranded into Good Part of Town and, soon afterwards, permanently discontinued. SaferRoute and Road Buddy, too, have ceased to exist. Fuelled by the extensive media coverage, SketchFactor enjoyed a brief spell of popularity, becoming the third-most downloaded navigation app, after Google Maps and Waze. However, this turned out to be a mixed blessing for the app's commercial success, as users flooded the platform with ironic reports warning pedestrians against the pretentiousness of hipster neighbourhoods, or the app's racism. In 2015, the app creators gave up on the project and turned SketchFactor into Walc, an app dedicated to making city streets more walkable (Marantz 2015).

REFERENCES

AFP. (2016, March 1). Lost Israel Troops Stray into Camp, Sparking Bloody Clashes. *Express.co.uk.* https://www.express.co.uk/videos/4781597764001/Lost-Israel-troops-stray-into-camp-sparking-bloody-clashes. Accessed 14 September 2020.

Agnew, J. A. (2003). *Geopolitics: Re-visioning World Politics.* London: Psychology Press.

Amoore, L. (2009). Algorithmic War: Everyday Geographies of the War on Terror. *Antipode, 41*(1), 49–69. https://doi.org/10.1111/j.1467-8330.2008.00655.x.

Badger, E. (2013, September 4). Enough Already with the Avoid-The-Ghetto Apps. *CityLab.* http://www.theatlanticcities.com/neighborhoods/2013/09/enough-already-avoid-ghetto-apps/6776/. Accessed 14 December 2017.

Bauman, Z. (1998). Space Wars: A Career Report. In *Globalization: The Human Consequences.* New York: Columbia University Press.

Beaumont, P. (2016, March 1). Israeli Soldiers' App use Leads to Deadly Fight in West Bank camp. *The Guardian*, online.

Berger, Y. (2018, March 6). Israeli Soldiers Filmed Doing Nothing as Settlers Attack Palestinians. *Haaretz.* https://www.haaretz.com/israel-news/israeli-soldiers-filmed-doing-nothing-as-settlers-attack-palestinians-1.5883745. Accessed 5 November 2018.

Berkow, B. (2012, January 9). Microsoft Patents 'Avoid Ghetto' GPS Feature. *Financial Post*. https://financialpost.com/technology/microsoft-pat ents-avoid-ghetto-gps-feature. Accessed 6 November 2018.

Bridle, J. (2018). *New Dark Age: Technology and the End of the Future*. London and Brooklyn, NY: Verso.

B'Tselem. (2013, August 26). Military Force That Killed Three Palestinians Stayed in Qalandia Camp Until Almost 7am. *B'Tselem*. https://www.btselem. org/press_releases/20130823_qalandia. Accessed 1 September 2020.

Byrne, M. (2015, March 22). The Simple, Elegant Algorithm That Makes Google Maps Possible. *Motherboard*. https://motherboard.vice.com/en_us/ article/4x3pp9/the-simple-elegant-algorithm-that-makes-google-maps-pos sible. Accessed 4 November 2018.

Capps, K. (2014, August 7). People Just Won't Give Up on Awful Neighborhood Apps. *CityLab*. http://www.citylab.com/tech/2014/08/peo ple-just-wont-give-up-on-awful-neighborhood-apps/375766/. Accessed 14 December 2017.

Carraro, V. (2019). Grounding the Digital: A Comparison of Waze 'Avoid Dangerous Areas' Feature in Jerusalem, Rio de Janeiro and the US. *GeoJournal*. https://doi.org/10.1007/s10708-019-10117-y.

Cashman, G. F. (2015, March 8). Iron Dome, WAZE Developers Among Israelis Chosen for 67th Independence Day Ceremony. *The Jerusalem Post | JPost.com*. https://www.jpost.com/israel-news/iron-dome-waze-develo pers-among-israelis-chosen-for-independence-day-torch-lighting-ceremony-393259. Accessed 22 September 2020.

Chandler, A. (2015, February 23). Jerusalem's Mayor Subdues Knife-Wielding Attacker. *The Atlantic*. https://www.theatlantic.com/international/archive/ 2015/02/jerusalem-mayor-Nir-Barkat-subdues-kife-wielding-attacker/385 795/. Accessed 4 November 2018.

Crampton, J. W. (2010). Cartographic Calculations of Territory. *Progress in Human Geography*. http://phg.sagepub.com/content/early/2010/01/28/ 0309132509358474.abstract. Accessed 24 October 2015.

Dumper, M. (2014). *Jerusalem Unbound: Geography, History, and the Future of the Holy City*. New York: Columbia University Press.

Dvir, N. (2014, September 12). Stones Thrown Against a Family That Entered Wadi Joz Instead of the Western Wall Because of a Navigation Mistake. *Ynet*. http://www.ynet.co.il/articles/0,7340,L-4570236,00.html. Accessed 7 December 2017.

Edney, M. H. (1997). *Mapping an Empire: The Geographical Construction of British India, 1765–1843*. Chicago, IL: University of Chicago Press. http://public.eblib.com/choice/publicfullrecord.aspx?p=3038725. Accessed 13 September 2016.

Elden, S. (2007). Governmentality, Calculation, Territory. *Environment and Planning D: Society and Space, 25*(3), 562–580. https://doi.org/10.1068/d428t.

Erakat, N. (2020, August 7). Extrajudicial Executions from the U.S to Palestine. *Just Security.* https://www.justsecurity.org/71901/extrajudicial-executions-from-the-united-states-to-palestine/. Accessed 24 September 2020.

Farías, I., & Widmer, S. (2018). Ordinary Smart Cities. How Calculated Users, Professional Citizens, Technology Companies and City Administrations Engage in a More-Than-Digital Politics. *TECNOSCIENZA: Italian Journal of Science & Technology Studies, 8*(2), 43–60.

Finkler, K. (2016, January 3). Lost Soldiers' Quick Reactions Saved Their Lives. *Israel National News.* http://www.israelnationalnews.com/News/News.aspx/208743. Accessed 16 November 2018.

Gekker, A. (2016, March 2). Waze, Wars, Wanton Disregard for User Privacy. *Casual Space.* http://alexgekker.com/blog/2016/03/02/waze-wars-wanton-disregard-for-user-privacy/. Accessed 22 May 2017.

Gross, J. A. (2020, February 13). The IDF's New Plan: From 'Waze of War' to a General Charged with Countering Iran. https://www.timesofisrael.com/from-waze-of-war-to-a-general-devoted-to-countering-iran-the-idfs-new-plan/. Accessed 24 September 2020.

Haggerty, K. D., & Ericson, R. V. (2000). The Surveillant Assemblage. *The British Journal of Sociology, 51*(4), 605–622. https://doi.org/10.1080/00071310020015280.

Har-Zahav, H. (2016, February 29). A Few Months Ago, When the Current Intifada Began... *Facebook.* https://www.facebook.com/haimhz/posts/10153927372701505. Accessed 16 December 2017.

Hassan, B. Y. (2016, March 11). Aspiring Palestinian Journalist Killed Months Before Graduation. *Arab America.* https://www.arabamerica.com/aspiring-palestinian-journalist-killed-months-before-graduation/. Accessed 14 September 2020.

Holmes, D. (2013, September 3). Pando: GhettoTracker Is the Worst Site on the Internet. *Pando News.* https://pando.com/2013/09/03/ghettotracker-is-the-worst-site-on-the-internet/. Accessed 14 December 2017.

Katzowitz, J. (2016, March 1). Israeli Soldiers, Armed Palestinians Engage in Firefight After Waze Mixup. *The Daily Dot.* https://www.dailydot.com/layer8/waze-error-israel-palestine-firefight/. Accessed 7 December 2017.

Kelley, R. D. G. (2019). From the River to the Sea to Every Mountain Top: Solidarity as Worldmaking. *Journal of Palestine Studies, 48*(4), 69–91. https://doi.org/10.1525/jps.2019.48.4.69.

Leszczynski, A. (2016). Speculative Futures: Cities, Data, and Governance Beyond Smart Urbanism. *Environment and Planning A, 48*(9), 1691–1708. https://doi.org/10.1177/0308518X16651445.

Marantz, A. (2015, July 29). When an App Is Called Racist. *The New Yorker.* https://www.newyorker.com/business/currency/what-to-do-when-your-app-is-racist. Accessed 30 November 2017.

Milo, P. (2012, January 6). *Microsoft Patents 'Avoid Ghetto' Feature for GPS Devices.* https://seattle.cbslocal.com/2012/01/06/microsoft-patents-avoid-ghetto-feature-for-gps-devices/. Accessed 6 November 2018.

OCHA. (2018, October 8). Over 700 Road Obstacles Control Palestinian Movement Within the West Bank. *United Nations Office for the Coordination of Humanitarian Affairs—Occupied Palestinian Territory.* https://www.ochaopt.org/content/over-700-road-obstacles-control-palestinian-movement-within-west-bank. Accessed 24 September 2020.

Orpaz, I. (2015, April 3). The IDF's New Boot Camp for Startup Entrepreneurs. *Haaretz.* https://www.haaretz.com/israel-news/.premium-1. 650383. Accessed 5 December 2017.

Peace Now. (2016). Data | Population. *Peace Now.* http://peacenow.org.il/en/settlements-watch/settlements-data/population. Accessed 15 December 2017.

Pow, C. P. (2015). Urban Dystopia and Epistemologies of Hope. *Progress in Human Geography, 39*(4), 464–485. https://doi.org/10.1177/030913251 4544805.

Reed, J. (2017, August 16). The West Bank's Dearth of Data Proves Difficult to Navigate. *Financial Times.* https://www.ft.com/content/7058f034-81d4-11e7-a4ce-15b2513cb3ff. Accessed 10 August 2018.

Ryan, S. (1996). *The Cartographic Eye: How Explorers Saw Australia.* Cambridge: Cambridge University Press.

Sadowski, J., & Pasquale, F. A. (2015). *The Spectrum of Control: A Social Theory of the Smart City* (SSRN Scholarly Paper No. ID 2653860). Rochester, NY: Social Science Research Network. https://papers.ssrn.com/abstract=265 3860. Accessed 7 May 2018.

Salamanca, O. J. (2016). Assembling the Fabric of Life: When Settler Colonialism Becomes Development. *Journal of Palestine Studies, 45*(4), 64–80. https://doi.org/10.1525/jps.2016.45.4.64.

Schechter, A. (2014, September 19). Guided by Waze into the Heart of the Palestinian-Israeli Conflict. *Haaretz.* http://www.haaretz.com/israel-news/.premium-1.616596. Accessed 22 May 2017.

Shelton, T., Zook, M., & Wiig, A. (2015). The 'Actually Existing Smart City'. *Cambridge Journal of Regions, Economy and Society, 8*(1), 13–25. https://doi.org/10.1093/cjres/rsu026.

Silver, J. (2013, October 2). Is Your Turn-By-Turn Navigation Application Racist? *American Civil Liberties Union.* https://www.aclu.org/blog/national-security/your-turn-turn-navigation-application-racist. Accessed 14 December 2017.

Stuart, H. (2016, October 5). Waze Lets Israelis Avoid Palestinian Areas, but Not the Other Way Around. *Motherboard*. https://motherboard.vice.com/en_us/article/jpgbg7/waze-lets-jewish-israelis-avoid-palestinian-areas-but-not-the-other-way-around. Accessed 7 December 2017.

Tashev, I. J., Couckuyt, J. D., Black, N. W., Krumm, J. C., Panabaker, R., & Seltzer, M. L. (2012, January 3). *United States Patent: 8090532—Pedestrian Route Production*. http://patft.uspto.gov/netacgi/nph-Parser?Sect1=PTO1&Sect2=HITOFF&d=PALL&p=1&u=%2Fnetahtml%2FPTO%2Fsrchnum.htm&r=1&f=G&l=50&s1=8,090,532.PN.&OS=PN/8,090,532&RS=PN/8,090,532. Accessed 6 November 2018.

Tawil-Souri, H. (2012). Uneven Borders, Coloured (Im)Mobilities: ID Cards in Palestine/Israel. *Geopolitics, 17*(1), 153–176. https://doi.org/10.1080/14650045.2011.562944.

Thatcher, J. (2013). Avoiding the Ghetto Through Hope and Fear: An Analysis of Immanent Technology Using Ideal Types. *GeoJournal, 78*(6), 967–980. https://doi.org/10.1007/s10708-013-9491-0.

Thatcher, J. (2017). You Are Where You Go, the Commodification of Daily Life Through 'Location'. *Environment and Planning A: Economy and Space, 49*(12), 2702–2717. https://doi.org/10.1177/0308518X17730580.

The New Arab. (2018, February 18). Israeli Settlers Violently Attack Palestinian Bus Driver Near Hebron. *alaraby*. https://www.alaraby.co.uk/english/news/2018/2/18/israeli-settlers-violently-attack-palestinian-bus-driver-near-hebron. Accessed 5 November 2018.

UNRWA. (2015, March). Kalandia Camp Profile. *UNRWA*. http://www.unrwa.org/sites/default/files/kalandia_refugee_camp.pdf. Accessed 1 September 2020.

Veracini, L. (2013). The Other Shift: Settler Colonialism, Israel, and the Occupation. *Journal of Palestine Studies, 42*(2), 26–42.

Waze. (2016). *Driver Satisfaction Index 2016* (p. 24). Waze. https://inbox-static.waze.com/driverindex.pdf. Accessed 18 November 2017.

Yanovsky, R. (2015, August 8). Waze Directing Users Away from 'PA Controlled' East Jerusalem. *Ynetnews*. http://www.ynetnews.com/articles/0,7340,L-4695380,00.html. Accessed 15 December 2017.

CHAPTER 4

A Glitch in Google Maps

Abstract In this chapter, I consider how Google assembles its maps from many sources, using algorithms and, occasionally, a set of loose norms, to decide which information to foreground for which users. This process goes largely unnoticed, but glitches such as the 2016 erasure of Palestine from Google Maps can cause users to notice and question Google's cartographic choices. Furthermore, in this case, the 'gap in the map' was taken up on Twitter by Palestinians abroad, allowing the performance of a national public outside the Israel/Palestine territory, in a curious twist of the historical relation between cartography and state-building. This episode demonstrates that maps continue to play an important role in supporting national imaginaries, albeit in distinctively postmodern ways.

Keywords Google · Twitter · Publics · Nation-building · Glitch · Palestine

In late July 2016, Palestine disappeared from Google Maps. Or, to be precise, the labels for the West Bank and the Gaza Strip vanished. The news circulated among Arab users via Twitter, with the hashtag *#ElQudsPalestinesCapital* ('Al-Quds', or, depending on the transliteration system, 'El Quds', is the Arabic name for Jerusalem). The Gaza-based Forum of Palestinian Journalists (2016) issued a press statement, urging

V. Carraro, *Jerusalem Online*, The Contemporary City, https://doi.org/10.1007/978-981-16-3314-0_4

Google to backtrack and apologise to the Palestinian people, framing the deletion as the result of pressures from Zionist lobbies. It also called upon Palestinian, Arab and international activists to protest Google's decision through a solidarity campaign and boycott. The statement helped to spread the news of Palestine's erasure among English-speaking news outlets. Prominent international newspapers including *The Guardian*, *The Washington Post* and the *New York Times* (Cresci 2016; Dewey 2016; Stack 2016) reported on the incident. New hashtags, such as *#BoycottGoogle*, *#ShameOnGoogleMaps*, *#PalestineIsHere* and *#NoPalestineNoGoogle*, appeared beside or in place of *#ElqudsPalestinesCapital*. For a few hours, the erasure of Palestine from Google Maps was all over the Internet. My inbox, too, filled with emails from friends and colleagues sharing commentaries about this questionable decision on the part of Google. Responding to requests from the press, Google eventually clarified that the name 'Palestine' had never appeared on Google Maps but that a glitch had caused some of the labels to temporarily disappear. This explanation quickly defused the online outrage for Google's actions, and the controversy died down as quickly as it had mounted.

From a media theory perspective, a glitch is 'a break from (one of) the protocolized data flows within a technological system' (Menkman 2011, p. 26), i.e. an interruption in the transmission of information. Artists have long used glitches in their practice, intentionally corrupting or manipulating data to create glitch art,[1] showing that, although defined by a breakdown of communication flows, glitches can carry meaning. In the *Glitch Feminism Manifesto*, curator and writer Legacy Russel (2012) conceptualises the glitch as at once a mistake in and a correction to the faulty system in which we all exist:

> (...) an error in a social system that has already been disturbed by economic, racial, social, sexual, and cultural stratification and the imperialist wrecking-ball of globalization—processes that continue to enact violence on all bodies—may not, in fact, be an error at all, but rather a much-needed erratum.

---

[1] Len Lye's short film *A Colour Box* (1935) is often regarded as the earliest example of glitch art. Produced as a commercial for the General Post Office, the work was created by painting abstract designs directly onto the film strip and combining these visuals with a Creole jazz soundtrack. At the 1936 Venice Film Festival, Nazi spectators disrupted the screening of what they saw as a glaring example of 'degenerate art' (Smythe 2013), an episode that I choose to see as an indication of glitches' unassuming subversive potential.

This glitch is 'a positive departure' from the machine, yet this departure originates within the machine itself, marked by liminality rather than subalternity. The glitch (plausibly from the Yiddish 'glitch', meaning slippery area) expresses a rejection of the dualistic definition of an inside and an outside, of two separate spheres, the digital and IRL ('in real life') world. 'The first step to subverting a system', elaborates Russel, 'is accepting that that system will remain in place; that said, the glitch says fuck your systems! Your delineations!'.

Articulated like this, the glitch creates an opportunity for reconsidering the (non-)representation of Palestine on Google Maps. As a computer error, it cannot be said to express an ideology, or a counter-narrative. However, it acquires meaning in relation to long-standing narratives about Palestine, and its existence or lack thereof, narratives that are routinely disseminated through the Google Maps machine. For a few weeks in 2016, the glitch 'corrected' this story, highlighting that Palestine is still an open question, whose existence and spatial definition are far from settled.

THE GOOGLE MAPS MACHINE

Google Maps is a daily presence in many people's lives, mine included. I use it to plan routes in unfamiliar places, to look up the nearest supermarket, or the capital city of Suriname. Generally, I trust it. Only when its machine fails to behave as expected do I stop to wonder how it works. I suspect the glitch has a similar effect on many users, making them question how Google decides what appears on its maps in a 'normal' situation. Let us consider Jerusalem as an example (see Fig. 4.1). Google collates information from a range of sources, and organises them on the page. On the right, the map portion of the screen draws on data by two private providers, Mapa GISrael and OrionMe, in addition to Google's own data. Mapa GISrael is an Israeli company based in Tel Aviv, while Orion-Me operates throughout the Middle East and North Africa and has offices in the United Arab Emirates and Lebanon. On the left half of the screen, an info bar complements the map with photos, text and links. The photos (Fig. 4.1, #1 and #4) come from Google Album Archive, the database where Google stores all the images shared across its platforms: Picasa, Google+, Blogger, Hangouts, Google Photos and Google Drive. The choice of which photos to display is automated, and we can only speculate about the selection criteria: relevant geo-tags, popularity among

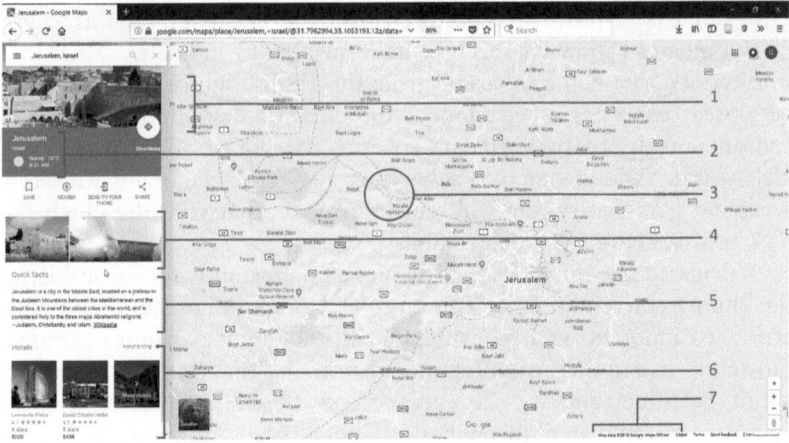

Fig. 4.1 Google Maps search results for 'Jerusalem', as of September 2018 (*Source* Google Maps, annotations in red by the author; Google and the Google logo are registered trademarks of Google LLC, used with permission)

users or across the web, image quality or the date of capture. Between these images, Google Maps displays data (Fig. 4.1, #2) pulled by other sources: weather information from weather.com and, more relevantly for this research, the city's location from Wikidata, the structured database that supports many Wiki projects, for example Wikipedia. Underneath, we find a 'Quick Facts' snippet (Fig. 4.1, #5), pulled from the relevant Wikipedia entry and, finally, links to Jerusalem hotels (Fig. 4.1, #6), with ratings and prices.[2]

Obviously, this is not the work of a cartographer, sitting in a dusty office in Palo Alto, regretting her degree while diligently copy pasting Wikipedia entries into Google Maps. To understand how this page comes

---

[2] The display of these ads is regulated by Google AdWords. Hotel managers bid on key words such as 'Jerusalem', 'budget accommodation' or 'hotel with pool'. Every time a user performs a search, Ad Words goes through the ad database in search of eligible ads that match it. The system then ranks these ads taking into account the bid amount, but also the relevance to the user search and the hotel's 'prominence'—which depends, among other things, on the hotel's Google review count and score, and its ranking in Google Search results (Google My Business Help 2018; Hotel Ads Center Help 2018). To make it harder for hotels to 'game' the algorithms, Google does not reveal exactly how these factors are weighted against one another.

together, it is useful to briefly consider how Wikidata works. Wikidata describes itself as 'a free and open knowledge base that can be read and edited by both humans and machines' (wikidata.org). Like Wikipedia, the project is organised in pages. Every subject covered by Wikidata is called an item, and every item has its own page. Each page collects the information related to the item in question, in the form of statements.[3] At a minimum, statements are composed of a property and a value, which can be complemented by additional values, qualifiers and references. Both items and properties are assigned a unique identifier, which is language-independent and, having no meaning, is unlikely to become obsolete or disputed. Such collections of terms and relations are called 'vocabularies' or 'ontologies'.

The 'Jerusalem' item has the unique identifier Q1218. At the time of writing, item Q1218 relates to approximately sixty statements. The property 'instance of' (unique identifier P31) is associated with the values: 'city', 'big city', 'capital' and 'city council'; the property 'coordinate location' (P625) with the values: '31°47'N' and '35°13'E'; the property 'image' (P18) with a photo of the Dome of the Rock; the property 'Instagram location ID' (P4173) with the value '213742495' and so on. Whenever a webpage needs some information about 'Jerusalem', it can do so by calling up the item identifier Q1218, and the identifiers for the relevant properties.

This way of organising knowledge, called semantic web, is increasingly prevalent, facilitating the exchange of information between machines without human intervention. For Graham (2015), the case of Jerusalem exemplifies why it is problematic. By presenting bits of data without context, Google passes a contested issue, i.e. that Jerusalem is in Israel, as an established fact. When knowledge about places is designed around machines rather than humans, Graham suggests, nuances and ambiguity are bound to get lost. Yet, I would counter, knowledge models designed around humans are not immune to similar transformations in meaning: the problem is inherent to the process of capturing reality through data. In fact, it is worth pointing out that the Wikidata model can record disputed and contradictory statements. Indeed, the statement about Jerusalem's location (the 'country' property) includes several

---

[3] In the interest of clarity and conciseness, my description of the Wikidata knowledge model does not go into details. Erxleben et al. (2014) provide a more comprehensive explanation.

clauses, specifying that Israel's claim on East Jerusalem are supported by the US, but rejected by the UN. Admittedly, Wikidata does not record that several countries[4] also challenge Israel's claims on West Jerusalem.

Wikidata is generated and maintained by users and 'bots' (a contraction of 'robots', referring to software that, as Wikidata page puts it, makes edits 'without the necessity of human decision-making') for users and machines. Clearly, automation is a key goal. Almost as important is an aspiration to standardise information. Indeed, Wikidata partially developed as a response to the inconsistencies between language-specific versions of the same Wikipedia entry. The 'Quick Facts' about Jerusalem, displayed in the Google Maps sidebar (Fig. 4.1, #5) offers a poignant example of these differences. The text captured in the screenshot comes from the Jerusalem Wikipedia entry in English, which reads:

> Jerusalem is a city in the Middle East, located on a plateau in the Judaean Mountains between the Mediterranean and the Dead Sea. It is one of the oldest cities in the world, and is considered holy to the three major Abrahamic religions—Judaism, Christianity, and Islam.

Inevitably, the text gives us a particular version of Jerusalem: for instance, it avoids mentioning either Palestine or Israel, describing the city location in relation to natural landscape features and emphasising its religious importance. This is not to say it is 'neutral', but rather that it is the result of a lengthy mediation process between English-speaking Wikipedia users and is thus shaped by that group's standards and values. When users search Google Maps in another language, they are presented with the corresponding Wikipedia entry, if one is available. So, for instance, as of June 2020, users searching in Hebrew read:

> Jerusalem is the capital of the State of Israel and the largest city in the country. As of 2019, there were approximately 927,000 residents. Jerusalem is home to the Israel government's institutions: the Knesset, the Supreme Court, the President's Residence, the Prime Minister's Residence and most ministries.

Here, Jerusalem is first and foremost the capital of Israel, a claim that, as we have seen, is widely disputed. The list of Israeli institutions located in

---

[4] Algeria, Lebanon, Iran, Indonesia and Pakistan are some examples.

the city serves to support that claim. The population data, an inevitably contentious issue in Jerusalem, is provided by an Israeli think tank. Meanwhile, users searching in Arabic are presented with:

> Jerusalem is the largest city in the occupied Palestinian territory and the most important religious and economic centre in the area. Other known Arabic names include bayt almuqdisi, alquds alsharifi, wa'uwlaa alqiblatayni; the Bible refers to it as Jerusalem, Israel officially calls it Yerushaláyim.[5]

In this version, Jerusalem is an occupied Palestinian city, and several names attest its Arab identity. Wikidata seeks to minimise the divergence in local perspectives by boiling down knowledge into precise, 'factual' statements. Through its clause structure, it delimits those statements, specifying that they only apply to part of the city, or that they are disputed by some parties.

In their book 'Objectivity', historians of science Daston and Galison (2007) argue that, throughout history, science has relied on different 'epistemic virtues', i.e. norms that have governed how science is produced how legitimate knowledge is produced. New virtues emerge locally and become dominant, without completely erasing older ones. If objectivity is, as Daston and Galison suggest, the aspiration to 'knowledge that bears no trace of the knower' (2007, p. 17), then it is at best a secondary virtue for Wikidata. Instead, the project seeks to minimise the room for interpretation and disagreement by breaking down knowledge into discrete claims by reputable actors. Wikidata's information is not necessarily accurate or fair, but it is open and contestable. Crucially, however, Wikidata is a repository for other web projects: it is not meant to be a resource for humans. Most people will access this information through services like Google, which select which statements, or parts of a statement, to include, thereby simplifying and transforming that information.

In most cases, Google Maps merges these disparate sources (private providers like Mapa GISrael and OrionMe for the map data, services like Wikidata and weather.com for the information in the sidebar) without intervening. However, for 'geopolitically sensitive regions' Google has developed a special strategy, explained in the company's *Public Policy* blog (Boorstin 2009; McLaughlin 2008). More than ten years after the

---

[5] Automatic translation of Hebrew and Arabic by Google, with minor edits for readability.

publication of these posts, the three principles outlined there remain relevant. First, the company seeks to represent 'ground truth' as objectively as possible, 'in consistency with Google's mission to organise the world's information and make it universally accessible and useful' (Boorstin 2009). Second, Google takes 'guidance from data providers that most accurately describe borders in treaties and other authoritative standards bodies' (Boorstin 2009), such as the UN Cartographic Section[6] or the International Standards Organisation (ISO). Importantly, through the word 'guidance', the phrasing of this post emphasises that Google's decisions take these sources into account but are fundamentally independent. As the blog explains, this is not only because such datasets are often insufficiently detailed and up to date, but also because Google does not see them as 'neutral', given that not all states are members (McLaughlin 2008). This is certainly true, and for that matter, member states have vastly different influence on the decisions that are being taken; but it does not follow that Google, a privately owned US company, should claim a greater degree of neutrality. Finally, Google seeks to 'localise the user experience while striving to keep all points of view easily discoverable in [their] products'. To this end, it uses 'primary, common, local' place names (Boorstin 2009). Here, 'primary' means that nicknames or local variations are not necessarily included. 'Common' suggests that the rule covers names in widespread use and not, for example, 'arbitrary' renaming by local rulers. Finally, 'local' means that Google Maps seeks to reflect the names used in the countries involved in the dispute (Boorstin 2009). In practical terms, this third principle means that the map is adjusted to match the expectations of local users, as well as national legislation regulating cartographic representations. So, for example, the Indian version of Google Maps shows the disputed regions at the borders with Pakistan and China as Indian, while users from the rest of the world are presented with a network of dashed grey lines, indicating disputed boundaries (Fig. 4.2).

Even if Google Maps does not use the name 'Palestine', when users input it in the search bar, they are directed to the expected location. The map (Fig. 4.3) shows three main labels: 'Israel', as the main country label in black, and 'Gaza Strip' and 'West Bank', the labels whose disappearance triggered the dispute, in grey as lower level regional labels, ostensibly part of Israel. The term 'Palestinian Territories', though in common

---

[6] This has since been renamed 'UN Geospatial Information Section'.

Map Localisation: Jammu and Kashmir Regions on Google Maps

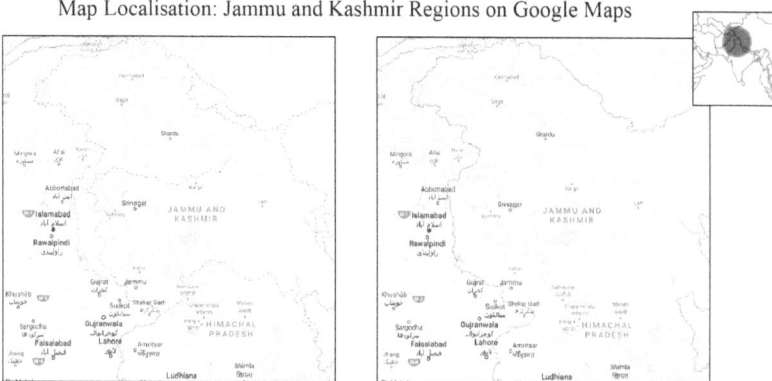

Localisation Settings: Global   Localisation Settings: India

Fig. 4.2 Comparison between the international and Indian versions of Google Maps, as of September 2018. The screenshot focuses on the regions of Jammu and Kashmir, where China, Pakistan and India have conflicting territorial claims. When search settings are set to 'global', Google Maps shows the borders as disputed; when search settings are set to 'India', the entire area is represented as part of India (*Source* Google Maps; Google and the Google logo are registered trademarks of Google LLC, used with permission)

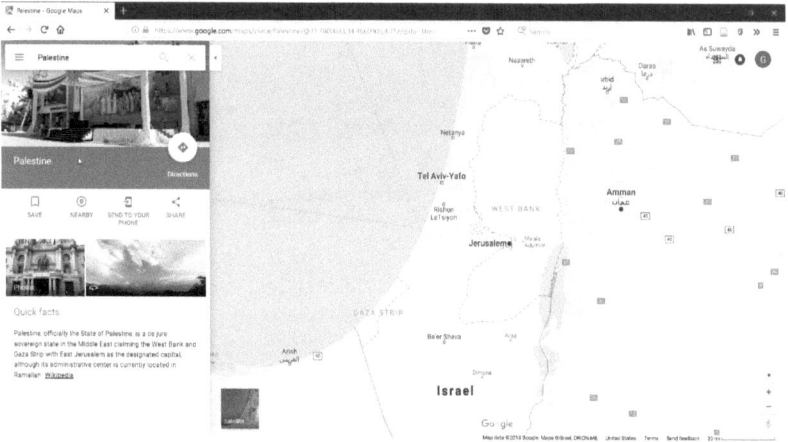

Fig. 4.3 Google Maps search results for 'Palestine', as of September 2018 (*Source* Google Maps; Google and the Google logo are registered trademarks of Google LLC, used with permission)

use, does not appear anywhere, in apparent contradiction with Google's second principle. Similarly, the map marks Jerusalem with a full black dot, thereby representing it as a country capital. This is arguably in line with Israeli local expectations, but in contrast with most authoritative sources, including the UN and most national governments, which consider Tel Aviv to be the capital, as discussed in Chapter 1. The borders separating the 1948 territories from the areas occupied in 1967 are marked by a dashed grey line and follow the armistice line recognised as the legitimate border by the U.N. and most international bodies. Incidentally, Google makes a patent exception to its first rule when it omits the barrier that, while running largely parallel to the armistice line, deviates from it to annex parts of the West Bank, including East Jerusalem. Presumably, Google's intention is to avoid legitimising the barrier as a *de-facto* border by representing it on the map, but there is no denying that the barrier is a 'ground truth' with immediate consequences for map users (Fig. 4.4).

Google's rules are of course vague and potentially contradictory, leaving considerable room for manoeuvre within this framework. Indeed, the first and third principles seem bound to conflict, obliging Google to

Fig. 4.4 Google Street View imagery of Jericho Road, as of September 2018. The bottom half of the screen shows the same road in Google Maps view. The West Bank barrier, which blocks the road, is not reported on the map (*Source* Google Maps; Google and the Google logo are registered trademarks of Google LLC, used with permission)

balance a positivist understanding of geographic truth with a postmodern, market-oriented aspiration to cater to customers' desires. Commentators tend to emphasise this latter impulse, describing Google as an 'agnostic cartographer' that prefers to affirm divergent worldviews rather than commit to a single reality (Gravois 2010; Merel 2015), a characterisation that is in line with Google's self-description as a neutral organiser of the world's information. There is something paradoxical in the notion that Google—the first site most people go to for answers—should embrace agnosticism, which literally means ignorance. Suggestive as it is, this juxtaposition downplays Google's role in curating geographic information, customising results based on its users' browsing history, and arranging them on the screen through its interface.

Surrounding maps with images and additional information from a variety of sources is an established practice, rather than a geoweb-related innovation. In this, the Google Maps interface closely adheres to the layout of older paper maps: a main map at the centre, a sidebar with text and images, and a source box tucked in a corner. Wood and Fels (2008) term the ensemble of these additional elements 'paramap', analogous to the 'paratext' in semiotics. The paramap advertises the map's authority, shaping its reception by 'immobilising our perception on [the cartographer's] chosen ground' (Wood and Fels 2008, p. 194). In Google Maps, however, the paramap becomes dynamic, varying depending on the user's location and browsing history, the images uploaded by Google users, the edits by Wikipedia contributors and so on. And, while cartographers carefully choose what to include in a static paramap on a case-by-case basis, on Google Maps these decisions are mediated by algorithms that abide by general models.

Automation is almost inevitable, given the tremendous amount of geographic information handled by Google. Yet, problematically, it is mobilised to create the impression that the map is simply relating all available information about the places we search, when in fact a lot of work goes into selecting, formatting sorting and arranging these data. In Google's case, automation also makes the map-making processes extremely opaque, since Google's algorithms are secretive, protected by trade law. Even if the code were accessible, it would likely be too complex to be understood by anyone other than expert code developers (Ananny and Crawford 2016). This lack of transparency puts Google in the perfect position for dropping the 'the-algorithms-did-it-card' (Glaser and Oremus 2018), blaming its choice and mistakes on the software.

When the West Bank and Gaza disappeared from Google Maps, all that was lost were two labels, but the consequences of computer errors can be devastating: between 2010 and 2015 the financial company Wells Fargo wrongfully foreclosed 400 homes because of what was later recognised as a 'software glitch' (Kosman 2018). Clearly, the problem is not whether the technical glitch happened, but how to conceptualise the agency of algorithms and the responsibilities of their makers.

Rightly highlighting these problems, progressive critiques of Google and Google Maps tend to pit corporations and (mapping) software as working together against (human) users (Fuchs 2011; Shaw and Graham 2017; Zook and Graham 2007). In this view, Google's algorithms appear as instances of encoded ideology, working steadily to advance their makers' projects of domination. The remedy is opening the map 'black-box' (Bittner et al. 2013), making software-sorting practices more transparent and facilitating a critical evaluation of code's cultural and spatial politics. Such framing reflects a justified anxiety at the perspective that bits of code, which we do not understand and over which we have no control, should exert such a great influence. It assumes that ideology, as embedded in the map code, is the main obstacle to more democratic mappings. Appeals to opening the black box of mapping software evoke a sudden 'eureka' moment where the map is finally broken down into its components, all connections and regulating mechanisms become clear: users can finally clean them of their 'ideological coating', tweak them and arrange them to their taste. In other words, the black box metaphor implies that there are clear inputs and outputs, distinct from the map itself. Once the map is deconstructed, it will be possible to fix it. The interactions between territories, states, users, software developers, corporations and knowledge, however, are messier than this model implies. It is not possible to neatly distinguish between map, input and output. Users participate in the making of maps through usage and interpretation; in turn, they are changed by maps in subtle way. If the Google Maps glitch matters, it is largely because of how people engage with it.

ERRATUM: THE PALESTINIAN QUESTION, IN 140 CHARACTERS

Palestine has never appeared on Google Maps, nor on most authoritative maps produced after 1948. As seen in Chapters 1 and 2, the erasure of Palestine as a nation, expressed in the well-known trope of 'a land without people for a people without a land', has been integral to Zionist narratives about the birth of Israel. It would be a mistake, however, to assume that Palestinian non-existence is simply a convenient story about the past—effectively a means to whitewash history. This fiction continues to play a key role in legitimising the occupation of the 1967 territories in the eyes of the international community. As legal scholar Noura Erakat aptly explains, the Israeli state has drawn upon 'the fiction of Palestinian nonexistence' to argue that the occupation of the West Bank and Gaza Strip is legally exceptional (*sui generis*). Because of the presumed sovereign void in these territories, Israel has claimed that the Palestinian Territories are 'neither occupied nor unoccupied' (Erakat 2017, p. 19), subjecting them to military control, without adhering to the international legal regime that regulates the occupation of territories by foreign powers.

Understanding the importance of this narrative to the Israeli state project is key to understanding why a simple map error caused such an upheaval. While many people oppose the denial of Palestine's existence, its lack of cartographic recognition is hardly news material. The Google error provided an opportunity for mobilising public attention to the problem, albeit only for a few days. To borrow Marres' term (2009), we could say that the glitch worked as a 'device of engagement', i.e. an object capable of 'dramatising' an issue in front of a public, giving it an empirical quality.

To examine these debates, I consider their unfolding on Twitter. A distinctive trait of Twitter's algorithms is that they particularly value content able to draw together users who are not already connected (Bang Carslen 2016). In other words, Twitter seeks to identify and promote topics that are discussed across the platform, even if they are not the most popular in absolute terms. This facilitates the emergence of global publics united by an interest in the topic that is being discussed rather than social ties. In the case examined here, it is useful to distinguish different (but overlapping) strands within this public. Three hashtags born out of the dispute work well as key words here: *#ElQudsIsPalestinesCapital*, *#NoPalestineNoGoogle*, *#PalestineIsHere*. Considering that Jerusalem was not affected by the change in label, the popularity of the *#ElQudsPalestinesCapital* hashtag may seem surprising. I suspect the phrase was so successful because it foregrounds the specific definition of Palestine shared by this user group. By laying emphasis on Jerusalem, the hashtag rejects the equation of Palestine with the Palestinian Territories, delimited by the Green Line and excluding West Jerusalem. To underscore this rejection, many tweets paired the hashtag with a stylised map of Palestine as defined during the British Mandate. A common trope in these posts is the notion that Palestine exists in the hearts of Arabs and especially Palestinians from the diaspora, over which Google has no control. Other recurrent themes are the future liberation of Palestine, and the notion that God's justice will in the end prevail over Google. Images of Palestinian national symbols often accompany these posts: stylised maps of Palestine, as mentioned, but also the Dome of the Rock, the Palestinian flag, slings or photos of unarmed Palestinian youth challenging Israeli soldiers. All in all, these tweets, many of which are in Arabic, have strong nationalist tones and focus more on affirming solidarity with Palestine than on advocating for change in Google Maps.

The second strand involves a larger, more international audience, intervening to demand Palestine's inclusion on Google Maps. Users often direct their tweets *at* Google, either through direct mentions using '@google', or through hashtags like *#NoPalestineNoGoogle* or *#ShameOnGoogleMaps*. In doing so, they seek to put pressure on the company through 'public shaming' or threats of switching to another search engine. In their efforts to change the map, many of these posts point to 'objective' evidence of Palestine's existence, rather than appealing to a deep personal conviction. For example, a user posted photos of the Palestinian delegation at the then-ongoing Olympic Games in Rio de Janeiro, commenting:

> Intern'l law, Olympic committee say Palestine is a nation state. It's not on Google maps. Why? #NoPalestineNoGoogle

Many other tweets cite international law or the UN's position on the matter to make a similar point. By drawing on these sources, their authors advocate for a limited definition of Palestine, coinciding with the Palestinian Territories. Often, users include links to a petition posted on the website *change.org* the year before, with the title 'Google, Put Palestine On the Map!' The petition received a tremendous boost from the mediatic exposure following the Google glitch: in March, 25,000 people had signed it, on 9 August over 150,000 (Griffin 2016); by the end of the month, they were over 300,000 (Bogen 2016). The text accuses Google of undermining the campaign for Palestinian independence through its omission, and of 'making itself complicit in the Israeli government's ethnic cleansing of Palestine' (Martin 2015). Users are urged to call on Google 'to recognize Palestine in Google Maps, and to clearly designate and identify the Palestinian territories illegally occupied by Israel'. Here, too, Palestine is defined as the territories that are 'illegally occupied' by Israel, i.e. those acquired in 1967. The phrasing suggests support for a two states solution, posing as a goal Palestinian independence, rather than Palestinian liberation.

Finally, many individuals, political groups and organisations took the controversy as an opportunity to draw attention to Palestine-related issues beyond the map. Some examples of this trend are messages publicising the consequences of the Israeli siege on Gaza, protesting the violent behaviours of the Israeli army in the West Bank or promoting the BDS movement. While these phenomena have little to do with Google Maps, these posts are directed at Google, and include hashtags such as *#PalestineIsHere* and *#BoycottGoogle*. To put these tweets into context, it is worth noticing that this is not the first time that Palestinian activists and sympathisers have used Google as an amplifier for their political messages. In 2006, Palestinian refugees used Google Earth to crowdsource a map of Palestinian villages destroyed during the Nakba (for an excellent analysis of this episode see Quiquivix 2014). In 2011, when Google launched its Street View service in Israel, activists reportedly organised the display of protest banners along the routes taken by the Street View Cars, so that they would be immortalised in Google's imagery (Oded 2011). In 2013, a group of Palestinian hackers hijacked Google.ps; for a few hours, users were redirected to a page that urged Google to add Palestine to its maps through the following message (reported in Yaron and Kenan 2013):

> Uncle google we say hi from Palestine to remember you that the country in google map not called Israel. Its called Palestine.
> # Question: what would happens if we changed the country tittle of Isreal to Palestine in google maps !!!
> It would be revolution ..
> So listen to rihanna and be cool :P

Meanwhile, a popular t-shirt printed in Jerusalem uses Google's interface to suggest that when people say Israel, what they really mean is Palestine. Clearly, these examples vary in their level of militancy and sophistication. Taken together, however, they suggest that the use of Google as a platform to raise the profile of Palestinian issues is a long term, if uncoordinated, strategy.

These examples suggest that the public mobilised by the Google glitch does not correspond to a bounded political community—say 'the Palestinians', or even 'the international community'. Its members have little in common, aside from being interested in the issue at stake and, perhaps, using the same map. In fact, they do not even agree on what the issue at stake is: Palestine's existence as a nation, its recognition on Google Maps, or denouncing Israel's policies in the occupied territories. In liberal political theory, 'the nation' and 'the public' are regarded as essentially synonymous. In everyday language, the word 'public' has two other meanings: first, it can refer to an audience gathered in the same place, as in: 'the public applauded the performer'; second, it can designate to the virtual, temporary gathering of people around a text or cultural artefact, as in 'you, reader, are part of the public of this book' (Warner 2005/2010). Within a social totality (e.g. the liberal national public) exist infinite publics (in the latter sense of the word): indeed, the 'fiction' of a national public is made possible by the gathering of publics around newspapers, speeches, books, films and maps. Conversely, the public of a global map, like Google Maps, stretches across many national publics, composed by users and commentators with different national identities, scattered all over the world. Within this heterogenous public, however, there exists a more cohesive 'sub-public', tweeting under the hashtag #ElQudsIsPalestinesCapital. This group is made up of people who identify with Palestine, united by a commitment to affirm—rather than debate—the nation's existence.

Map-making and state-making are closely related processes: on the one hand, states have held an historical near-monopoly on cartography; on the other, maps have been instrumental to the creation of nation-states. Often, the process has been violent: as seen in Chapter 1, maps facilitated exploration, colonisation, war and territorial control. But maps also help to solidify such conquests through soft power, supporting the formation of cultural, social and political identities. Maps, like media, help strangers imagine themselves as part of the same territorially bounded imagined community, the same nation (Anderson 1983/2006). Palestine, however, has never been a sovereign state with full 'map-making powers'. An independent Palestinian cartography only developed starting from the 1970s, and it is only since the establishment of the PA in the 1990s that Palestinians were able to build a national map-making infrastructure (Leuenberger 2012; Quiquivix 2012). Today, the PA's mapping efforts are restricted to the West Bank and, even within those boundaries, limited territorial control poses serious challenges to map-makers (Bier 2017, pp. 106–120). Most importantly, as also discussed in Chapter 3, at present Palestine is a nation splintered into distinct political communities: the Palestinian citizens of the West Bank, itself broken up into an 'archipelago' of non-continuous land masses, Palestinian citizens in the besieged Gaza Strip, Palestinians with Israeli citizenship, Jerusalem permanent residents and the Palestinian diaspora. These groups are subjected to different governments, have different political rights and uneven access to Palestinian land: in short, their experiences of 'being Palestinian' are considerably different. Against this backdrop, social media provide important channels for enacting a transnational Palestinian identity (Aouragh 2012). Interpreted in this light, the Google glitch provided an opportunity for Palestinian users to perform a shared identity online. Somewhat paradoxically, then, the cartographic erasure of Palestine, taken up on Twitter, offers a platform for staging Palestine's existence, not only discursively, but also through the enactment of a Palestinian national public.

* * *

In this chapter, I have taken as a starting point the Google Maps glitch that in 2016 caused the disappearance of some map labels, leading users to protest the cartographic erasure of Palestine. I started by examining in detail how Google Maps is assembled by mashing up different sources through a process that is largely automated and obscure. This machine plays a crucial role in regulating what information is visible to whom, while making it difficult to hold any one person or institution accountable for what is represented. While, most of the time, the production, selection and sorting of information happens in the background, unnoticed by most, glitches render visible the workings of the Google Maps machine, as happened in the case discussed here. Google's error directed attention to the long-standing erasure of Palestine, providing an 'erratum' to the standard representation of Israel/Palestine, and an opportunity for a territorially scattered Palestinian public to gather around a national map, or rather the lack thereof.

Hopefully, it is clear that my intention is not to counter the critique of Google with a naively optimistic story about Twitter's democratic potential. The algorithms that govern both Google Maps and Twitter are designed with the interests of these companies in mind, which generally means the maximisation of user clicks, and thus the profits made from advertising. Following this logic, they determine who should see which posts and articles based on users' online behaviours, aggregating 'big data publics' (Harper 2017) that may or may not share a social identity. These mechanisms constrain citizens' capacity to determine which issues are worth publicising, since when attention is a prized commodity, it is the market that decides (Harsin 2015). However, algorithms are not all-powerful. Platforms need to accommodate local legislations, cartographic standards, user expectations, demands and 'hijacking' tactics, among other things. Every stage entails a process of translation: desired outcomes need to be expressed as code, users need to be categorised according to recorded behaviours, photos according to their metadata, etc. As a result, algorithms enact their creators' intentions, but never exactly (Introna 2011, p. 117), informing mapping in many different ways: they encode and order information, facilitate exchanges across borders and distances, aggregate publics and consumer pools, format the spaces for political engagements, and direct attention towards consolidated geographical 'facts'.

Does a glitch in Google Maps really matter, in the grand scheme of things? I find the metaphor of the glitch helpful for considering Google's error with curiosity but without exaggeration. It is tempting to either overstate the event, romanticising user reactions as acts of political resistance, or to dismiss it entirely as yet another example of user misinformation. When I reflect on these issues, I often think of the exchange I had with a Palestinian friend soon after the glitch happened. Her hope, she told me, was to transform the 'real' world, not Google. She certainly has a point: demands to 'put Palestine on the (Google) map' problematically reframe the problem as one of digital recognition, rather than liberation. The trope of the glitch, however, reminds us that Google and its users do not exist in a separate, digital world. An error on Google maps will not lead to radical political change, but it will likely inform how 'real' people think of Palestine in the long run. Transplanted from the arts into the social sciences, the trope of the glitch risks losing much of its nuance and playfulness. Yet it is a risk worth taking, because the possibilities contained in the glitch can only be realised when we take advantage of it to question the machine and change the story we are being told.

REFERENCES

Ananny, M., & Crawford, K. (2016, December). Seeing Without Knowing: Limitations of the Transparency Ideal and Its Application to Algorithmic Accountability. *New Media & Society*. https://doi.org/10.1177/146144481 6676645.

Anderson, B. R. O. (1983/2006). *Imagined Communities: Reflections on the Origin and Spread of Nationalism* (Rev. ed.). London and New York: Verso.

Aouragh, M. (2012). *Palestine Online: Transnationalism, the Internet and the Construction of Identity* (New ed.). London: I.B. Tauris.

Bang Carslen, H. (2016). The Public and Its Algorithms: Comparing and Experimenting with Calculated Publics. In L. Amoore & V. Piotukh (Eds.), *Algorithmic Life: Calculative Devices in the Age of Big Data*. London and New York: Routledge, Taylor & Francis Group.

Bier, J. (2017). *Mapping Israel, Mapping Palestine: How Occupied Landscapes Shape Scientific Knowledge*. Inside Technology. Cambridge, MA: The MIT Press.

Bittner, C., Glasze, G., & Turk, C. (2013). Tracing Contingencies: Analyzing the Political in Assemblages of Web 2.0 Cartographies. *GeoJournal, 78*(6), 935–948. https://doi.org/10.1007/s10708-013-9488-8.

Bogen, M. (2016, August 26). How Google Has Managed the Tricky Balancing Act of Geopolitics. *Newsweek*. https://www.newsweek.com/googles-place-geopolitics-493769.

Boorstin, B. (2009, April 12). When Sources Disagree: Borders and Place Names in Google Earth and Maps. *Google Public Policy Blog*. https://publicpolicy.googleblog.com/2009/12/when-sources-disagree-borders-and-place.html.

Cresci, E. (2016, August 10). Google Maps Accused of Deleting Palestine—But the Truth Is More Complicated. *The Guardian*. http://www.theguardian.com/technology/2016/aug/10/google-maps-accused-remove-palestine.

Daston, L., & Galisonm, P. (2007). *Objectivity*. New York and Cambridge, MA: Zone Books; Distributed by the MIT Press.

Dewey, C. (2016, August 9). Google Maps Did Not "Delete" Palestine—But It Does Impact How You See It. *Washington Post*, sec. The Intersect. https://www.washingtonpost.com/news/the-intersect/wp/2016/08/09/google-maps-did-not-delete-palestine-but-it-does-impact-how-you-see-it/.

Erakat, N. (2017). Taking the Land Without the People: The 1967 Story as Told by the Law. *Journal of Palestine Studies, 47*(1), 18–38. https://doi.org/10.1525/jps.2017.47.1.18.

Erxleben, F., Günther, M., Krötzsch, M., Mendez, J., & Vrandečić, D. (2014). Introducing Wikidata to the Linked Data Web. In P. Mika, T. Tudorache, A. Bernstein, Ch. Welty, C. Knoblock, D. Vrandečić, P. Groth, N. Noy, K. Janowicz, & C. Goble (Eds.), *The Semantic Web—ISWC 2014* (pp. 50–65). Lecture Notes in Computer Science. Springer International Publishing.

Fuchs, C. (2011). A Contribution to the Critique of the Political Economy of Google. *Fast Capitalism, 8*(1), 31–50.

Glaser, A., & Oremus, W. (2018). Embracing Deplorable Status. *If Then by Slate.* http://www.slate.com/articles/podcasts/if_then/2018/08/if_then_t alks_to_right_wing_media_expert_will_sommer_about_the_latest_far.html.

Google My Business Help. (2018). *Improve Your Local Ranking on Google.* https://support.google.com/business/answer/7091?hl=en.

Graham, M. (2015, November 30). Why Does Google Say Jerusalem Is the Capital of Israel? *Slate.* http://www.slate.com/articles/technology/future_ tense/2015/11/why_does_google_say_jerusalem_is_the_capital_of_israel. html.

Gravois, J. (2010, July 9). The Agnostic Cartographer. *Washington Monthly.* https://longreads.com/picks/the-agnostic-cartographer/.

Griffin, A. (2016, August 9). Everyone Thinks Palestine Was Removed from Google Maps. It's Far More Tragic than That. *The Independent.* http:// www.independent.co.uk/life-style/gadgets-and-tech/news/google-maps-rem oves-palestine-says-petition-but-the-truth-is-far-more-complicated-than-that- a7181296.html.

Harper, T. (2017). The Big Data Public and Its Problems: Big Data and the Structural Transformation of the Public Sphere. *New Media & Society, 19*(9), 1424–1439. https://doi.org/10.1177/1461444816642167.

Harsin, J. (2015). Regimes of Posttruth, Postpolitics, and Attention Economies. *Communication, Culture & Critique, 8*(2), 327–333. https://doi.org/10. 1111/cccr.12097.

Hotel Ads Center Help. (2018). *Frequently Asked Questions for Hotel Owners.* https://support.google.com/hotelprices/answer/7219055?hl=en.

Introna, L. D. (2011). The Enframing of Code: Agency, Originality and the Plagiarist. *Theory, Culture & Society, 28*(6), 113–141. https://doi.org/10. 1177/0263276411418131.

Kosman, J. (2018, August 6). Wells Fargo Admits Glitch Led to Hundreds of Home Foreclosures. *New York Post.* https://nypost.com/2018/08/05/ wells-fargo-admits-glitch-led-to-hundreds-of-home-foreclosures/.

Leuenberger, C. (2012, December). Mapping Israel/Palestine. *Bulletin Du Centre de Recherche Français à Jérusalem* (23). https://bcrfj.revues.org/ 6859.

Marres, N. (2009). Testing Powers of Engagement: Green Living Experiments, the Ontological Turn and the Undoability of Involvement. *European Journal of Social Theory, 12*(1), 117–133. https://doi.org/10.1177/136843100809 9647.

Martin, Z. (2015). Google, Inc: GOOGLE: Put Palestine On Your Maps! *Change.Org.* https://www.change.org/p/google-inc-google-put-pal estine-on-your-maps.

McLaughlin, A. (2008, August 4). How Google Determines the Names for Bodies of Water in Google Earth. *Google Public Policy Blog*. https://public policy.googleblog.com/2008/04/how-google-determines-names-for-bodies. html.

Menkman, R. (2011). *The Glitch Moment(Um)* (G. Lovink, Ed.). Network Notebook 4. Amsterdam: Institute of Network Cultures.

Merel, E. R. (2015). Google's World: The Impact of Agnostic Cartographers on the State-Dominated International Legal System. *Columbia Journal of Transnational Law*, *54*, 424.

Oded, Y. (2011, August 24). Protests Against Israeli Occupation Make It to Google Street View. *Haaretz*. https://www.haaretz.com/1.5156265.

Quiquivix, L. (2012). *The Political Mapping of Palestine*. Chapel Hill: University of North Carolina at Chapel Hill. https://cdr.lib.unc.edu/indexablecontent/ uuid:b079da59-f8ff-4ec5-b1e2-9e335bd059c0.

Quiquivix, L. (2014). Art of War, Art of Resistance: Palestinian Counter-Cartography on Google Earth. *Annals of the Association of American Geographers*, *104*(3), 444–459. https://doi.org/10.1080/00045608.2014. 892328.

Russel, L. (2012, December 10). Digital Dualism and the Glitch Feminism Manifesto. *Cyborgology*. https://thesocietypages.org/cyborgology/2012/12/10/ digital-dualism-and-the-glitch-feminism-manifesto/.

Shaw, J., & Graham, M. (2017, January). An Informational Right to the City? Code, Content, Control, and the Urbanization of Information. *Antipode*, *49*, 907–927. https://doi.org/10.1111/anti.12312.

Smythe, L. (2013, May). Len Lye: The Vital Body of Cinema. *October*, *144*, 73–91. https://doi.org/10.1162/OCTO_a_00141.

Stack, L. (2016, August 11). No, Google Says, It Did Not Delete "Palestine" from Its Maps. *The New York Times*, sec. Middle East. https://www.nytimes. com/2016/08/12/world/middleeast/google-palestine.html.

Warner, M. (2005/2010). Publics and Counterpublics. In *Publics and Counterpublics* (1. paperback ed., 3. print, 4. Print). New York, NY: Zone Books.

Wood, D., & Fels, J. (2008). The Natures of Maps: Cartographic Constructions of the Natural World. *Cartographica: The International Journal for Geographic Information and Geovisualization*, *43*(3), 189–202.

Yaron, O., & Kenan, I. (2013, August 27). Hackers Hijack Google Palestine Site with Anti-Israel Message. *Haaretz*. https://www.haaretz.com/.premium-hac kers-hijack-google-palestine-site-1.5325895.

Zook, M., & Graham, M. (2007). The Creative Reconstruction of the Internet: Google and the Privatization of Cyberspace and DigiPlace. *Geoforum*, Theme Issue: Geographies of Generosity, *38*(6), 1322–1343. https://doi.org/10. 1016/j.geoforum.2007.05.004.

CHAPTER 5

Naming Jerusalem on OpenStreetMap

Abstract In this chapter, I discuss the politics of toponomy in Jerusalem on OpenStreetMap, a collaborative mapping platform that, despite being ostensibly open to contributions from everyone, offers a surprisingly one-sided representation of Jerusalem. The absence of Palestinian names from the map could be interpreted as an objective reflection of reality 'on the ground', or a sign that Palestinian views continue to be excluded and erased. Instead, drawing on my experience working for Grassroots Jerusalem and on an analysis of historical sources, I argue that they point to ongoing confrontations over both the purpose of mapping and the city's identity. Through its mapping standards, OpenStreetMap intervenes on these confrontations in ways that tend to consolidate the municipality's naming power, but also to show its limits.

Keywords OpenStreetMap · Crowdsourcing · Jerusalem · Street names · Critical toponymies

OpenStreetMap, or OSM, is frequently described as 'the Wikipedia of maps': an open platform that enables users to freely access and download geographic data, add to it, make edits, raise objections and debate how to best represent a certain place. It may not be as popular as Google Maps or Waze, yet nearly seven million users (as of October 2020) make

V. Carraro, *Jerusalem Online*, The Contemporary City, https://doi.org/10.1007/978-981-16-3314-0_5

it a major player in the field. In addition, many companies, public institutions and non-governmental organisations use OSM data, with examples including Apple, Uber, the FBI and the Red Cross. Like Wikipedia, OSM is crowdsourced (meaning it draws on user contributions to generate content), non-profit and community-driven. Arguably, this kind of project represents the best that the Internet has to offer: a relatively democratic site where information can be shared. This information is generally of high quality, translated in several languages and, perhaps most importantly, open to contestation. Given the growing influence of corporations like Google on geographic information, and maps in particular, Internet scholar John Naughton (2019) has characterised this kind of project as 'a ray of light': a demonstration that, collectively, we can build an alternative to corporate models, navigate disagreement and come to 'an approximation of the truth'.

Searches for Jerusalem on the OSM website, however, return a strikingly one-sided perspective on the city. The map appears to be almost entirely in Hebrew. A thick purple line, indicating an international boundary, runs around it, reaffirming Israel's claims over it. There are few clues that any of the information is contentious: user comments expressing disagreement are rare and buried in old forum threads or in the database recording previous edits. These observations are substantiated by previous studies (Bittner 2014, 2016), which underline a significant imbalance in the representation of the city's demographic groups: secular Jewish neighbourhoods are mapped in far greater depth than ultra-orthodox and Arab areas. In the case of Jerusalem, then, OSM falls short of offering a radical alternative to the uneven mapping of commercial providers.

Such an imbalance suggests that Palestinians are hardly involved in OSM. To an extent, this does not come as a surprise. Empirical studies of both Wikipedia and OSM consistently show that, although everyone can participate in theory, in practice participation depends on resources that remain unavailable to large segments of the world population: an Internet connection, free time and relatively high levels of literacy and technical skills. Unsurprisingly, the distribution of these resources reflects long-standing patterns of inequality among genders, ethnic groups, the Global North and South, and rural and urban areas. In turn, who contributes to these projects shape what kind of information is recorded. For instance, in a study on the gendered nature of the geoweb, Monica Stephens (2013) found that the OSM database privileges male views on the environment, reflecting the project's predominantly male demographics—96%, according to an early survey (Haklay and Budhathoki 2010). Stephens found that OSM featured only two categories for childcare facilities— 'kindergarten' and, oddly, 'baby hatch'. By contrast, it classified venues for night entertainment predominantly attended by men in detail, distinguishing for example between 'bar', 'pub' and 'biergarten', or between 'night club', 'swinger club', 'strip club' and 'brothel'. In short, projects like Wikipedia may have broadened participation in the production of geographic knowledge, but only few, relatively privileged groups benefit from these transformations. The 'crowd' from which this information is 'sourced' is smaller and more homogenous than it is often assumed.

On the other hand, considering the generally high levels of education among Palestinians (UNDP 2014), and the fact that a significant proportion of Palestinians live in areas served by Israeli Internet providers, technical and linguistic barriers do not seem able to fully account for the near-absence of Palestinian perspectives on OSM. Instead, I argue that there are other factors at play, more specific to the Israel/Palestine context. This chapter expands on the analysis of under-representation in crowdsourced projects by looking beyond 'the crowd' that authors OSM, using as a starting point the dispute that surrounds map labels of Jerusalem. First, I consider the edit war surrounding the Jerusalem node, focusing on the perspectives of GJ. Their example shows that participation dynamics are more complex than is often assumed, and that 'opting-out' can be a form of political expression. In the second part of the chapter, I examine the interplay between OSM mapping standards and the ongoing confrontations around the city's street names. Through this approach, the distinction between map authors and mapped territories fades, pushing the scope of analysis beyond map contributors, to cartographic standards and protocols, practices of usage and interpretation, and crucially, to the places and objects that the map represents.

THE EDIT WAR OVER JERUSALEM

As OSM enthusiasts are keen to emphasise (e.g. Wroclawski 2014), one important difference between Wikipedia and OSM is that OSM users are less likely to engage in edit wars, i.e. instances when users repeatedly revert each other's edits. While there is no strict definition of how many edits are needed to start a war, and no monitoring system, it is true that mapping disputes do not happen often. Meanwhile, there are thousands of 'wars' on Wikipedia, often around topics as trivial as the origins of Caesar Salad or the spelling of the word 'yoghurt' (for more entertaining examples, see the Wikipedia entry on 'Wikipedia:Lamest edit wars'). Perhaps precisely because they are so rare, OSM edit wars often become notorious, at least in the small circles of OSM users and researchers. This is also the case for the dispute that arose starting from 2009 around the Jerusalem node, to an extent that it has become a canonical example of edit war in the OSM academic literature (Bittner 2016; Perkins 2014).

To understand this episode, an explanation of the OSM data structure is in order. OSM represents any object through a node, line or polygon, to which is associated a series of tags and values. So, the node representing Jerusalem has a 'place' tag equal to 'city', distinguishing it from other points on the map representing landmarks, bus stops or shops. Its 'population' tag records a value of '780200', its 'is_in:continent' tag is designated as 'Asia', and so on. The primary 'name' tag contains each feature's name in the local language. Mappers can record a place name in other languages, storing this information in separate tags, such as name:en for the English name or name:zh for Mandarin, etc. When cartographers and developers want to use this data to create a map, they can select which ones to include in their visualisation with code, as they do with Wikidata (see Chapter 4). The process of translating the data model into a map is called rendering. The OSM website also includes a rendered 'slippy map' displaying selected features in a standard graphic style; this is the map I described at the beginning of this chapter. By default, this map shows only the main name tag, as do many services that rely on the same rendering rules.

The Jerusalem node first appeared on the map in 2007, with a 'name' tag set to 'Jerusalem' (in English), and a 'is_in' tag set to 'Israel'. Between 2009 and 2011, users changed the node's name back and forth a dozen times, alternating between Hebrew—'Yerushalayim' (ירושלים)—and a combination of Hebrew and either English—Jerusalem—or Arabic—'Al Quds' (القدس). One of the often-overlooked peculiarities of this 'edit war' is that when the debate mounted in the OSM user forums, in the summer of 2011, the belligerent editing activities were already winding down, and Jerusalem had been continuously labelled in Hebrew for six months.

At the time, GJ had just started its activities and had no mapping experience. They were directed to OSM by Mikel Maron, a veteran OSM user and member of the OSM Foundation. Maron had been involved in several initiatives aimed at jump-starting or fostering local mapping communities in Nairobi, Dar es Salaam, Kampala and Ramallah, as reported in his blog *brainoff.com*. He provided the GJ staff and volunteers with basic mapping training, including how to use GPS units to record points on a map, and how to upload them onto OSM. GJ was interested in using maps to represent Jerusalem from a Palestinian perspective, but thought Palestinian Jerusalemites would refuse to contribute to a map that, when accessed from Jerusalem, is in Hebrew. The labelling of Jerusalem as Yerushalayim, in particular, was likely to alienate many potential mappers. Maron offered to help organise a meeting with the Israeli OSM mappers, with the aim of discussing how to edit the map in a way that would encourage Palestinians to participate, yet still be acceptable to Israeli contributors.

The meeting ended without a definitive solution, but the following proposal was put forward: there would be two nodes, one in Hebrew and one in Arabic, and each 'side' would decide on the appropriate tags. This arrangement was to be discussed with other mappers, and, if approved, implemented. In the meantime, the OSM Data Working Group (DWG), a team of volunteer expert mappers responsible for mediating and settling mapping disputes, removed the existing name tag from the Jerusalem node, likewise effectively erasing Jerusalem from the map. It also added a note that urged users to refrain from making any further edits until the involved parties agreed on a compromise.

After the meeting, the controversy unfolded on a forum thread titled '"Edit War" over Jerusalem - meeting with East Jerusalem mappers', where the debate largely revolved around establishing who had the right to intervene on the node, and through what means. To begin with, the Israeli mappers took issue with how OSM, here represented by Maron and by the DWG, had dealt with the dispute. The DWG, they insisted, had overstepped its authority, needlessly deleting a tag that had not been edited for months and could therefore not be considered disputed. Clearly, given the city's political and spiritual importance, its erasure from the map acquired a strong symbolic meaning.

To make matters worse, the issue had been raised not by 'true' OSM mappers, but by outsiders with political motivations (for more on the perspectives of Israeli mappers see Bittner 2016). In this view, political activists make for poor mappers because they are unable or unwilling to differentiate objective data from their political views:

The problem here seems to be that the people that would like to create new facts are caring only political and do push the correct buttons.

On the other side, you see israeli power osm mappers, that do care about correct data without taking too much care about a rendering engine. And we do not that much care about politics.[1] (Mr_Israel 2011)

Users contributing to the discussion repeatedly stressed this juxtaposition, insisting that their position was based on factual observations, and thus not political. For example, a 'two nodes solution' was deemed unacceptable since '*physically* there is no border between east and west parts of Jerusalem, and for all **practical** purpose Jerusalem is one city, controlled by Israel'. By contrast, Palestinians had still to demonstrate their integrity as mappers and, crucially, their good sense:

> Why should they be granted the prominent label of Al-Quds right from the start? Let them first map it, let them show that they care about this map, and then they will be on equal ground with us to enter into discussion about naming/rendering/whatever. Let them show they can make compromises.

Particularly controversial was the involvement of Micha, a Jewish-Israeli member of GJ known to the Israeli public for his anti-Zionist activism. The Israeli OSM mappers had been prepared to negotiate with their Palestinian counterparts, but they read the involvement of 'so-called mappers' with a radical political agenda as an attempt to 'hijack' the mapping platform. Clearly, the dispute around the node is as much about the city's 'real' name as it is about what it means to *map*, and who has the right to do it. The users involved in this discussion share a positivist understanding of mapping as the process of collecting objective information about the external world in an organised database. Legitimate mappers should subscribe to this view. They should have adequate skills and experience, or at least aspire to acquire them. They should also be committed to OSM, putting the project's growth and success above any personal or political agenda. As newbies intentioned to use mapping as an advocacy tool, GJ embodied everything mappers should *not* be.

---

[1] All the quotes by Israeli OSM mappers in this section are taken from the OSM forum thread '"Edit War" over Jerusalem - meeting with East Jerusalem mappers' (July 2011).

Pointing at the platform's definition of dispute, the Israeli OSM users insisted that, since no 'real' mapper was contesting the node's tag, there was no controversy to speak of, or as one of the users emphatically put it, 'there was never a osm editing war! never!' Following the same rationale, several years later, a representative of the OSM DWG responded to my inquiries confirming that the dispute was 'settled a long time ago', since there is no evidence of repeated edits to the node. Yet, it is worth noting here that the controversy was driven by a handful of users. Most of the edits to the Jerusalem node—8 out of 11—were made by only two accounts. Similarly, of the 55 posts that comprise the forum discussion examined above, 46 were authored by the same four users. There is no doubt that many more people, including thousands of Palestinian Jerusalemites, disagree with how OSM presents Jerusalem, yet they do not express their views within OSM, through edits or forum posts. If the edit war appears to be settled, it is not because a consensus has been reached, but because disagreement has become invisible. Thus, the questions of what counts as an edit war, who should be able to contest the map and when should we assume that a mapping 'war' has been resolved are part of what is being disputed.

One of the reasons the controversy dragged on is that the OSM DWG insisted any solution needed to be endorsed by 'both sides', yet Palestinians stopped participating after the initial meeting. As a result, their perspective is missing from the debates on the OSM forum and mailing list, and also from scholarly accounts of the dispute (most notably Bittner 2016). For some, such a retreat from participation could support the forum users' assessment that GJ are not serious mappers, since they lost interest so quickly. For others, it could illustrate how OSM technical barriers and internal hierarchy work to marginalise and ultimately exclude the voices of inexperienced mappers from social and ethnic minorities. In contrast, I advance a third interpretation, suggesting that GJ *chose* to stop engaging with OSM, and that this choice not to map was as political as mapping can be. To this end, I draw on interviews with the two GJ members who attended the meeting and on my experience with the NGO.

As mentioned previously in this chapter, GJ was a young NGO when the dispute erupted, interested in the tools for advocacy mapping offered by OSM. According to GJ mapping coordinator, Amany, the July 2011 meeting was an occasion to find out more about the platform: 'what is this tool, or what we are going to do with this tool, how is it efficient for us, or useful for us (...) We really didn't know where we were going, or what we wanted to do...' (personal communication, February 2016). Her main recollection of the notorious meeting was that GJ and the Israeli mappers could not agree on anything. Likely due to her lack of familiarity with OSM and its procedures, she did not know that the discussion continued in the online forums after the meeting. In any case, she told me, she doubts she would have participated, had she known about it:

> Debating with Israelis is really not my goal. In any forum (...) And, the maximum which I can get from Israelis is that I might write Ras al Amud, or Silwan, or East Jerusalem... which is OK because it does reflect the reality, but not one hundred percent. But it is not justice, and freedom for a Palestinian, I won't really be able to go back to Ein Karem and see refugees in Ein Karem in OSM, which is a destroyed village in the West part of Jerusalem. (Personal communication, February 2016)

From her point of view, even the most 'Palestinian-friendly' solution would involve recognising that West Jerusalem is unquestionably Israeli, a 'fact' that GJ rejects wholeheartedly. Thus, any compromise would necessarily favour the Israeli side, which has the upper hand, on OSM and offline. In this context, Palestinian participation would give legitimacy to what is, in GJ's eyes, an unjust map.

Micha, the Jewish-Israeli member of GJ who intervened in the forum discussion, expressed a similar sentiment in our interview, questioning whether participation was the best course of action in the circumstances:

> Well, that was six or seven years ago, and I didn't have much of an understanding of the principles of solidarity, the kind of "dharma" [a Hindu concept that refers to a universal moral code], that I now respect as a white Israeli being part of the pro-Palestinian movement. Since then, I developed an increased awareness of my own privilege, and if I had the chance to do it again, I would probably stand in solidarity with Amany, and also refuse to engage. (Personal communication, October 2017)

In explaining their choices regarding participation, they both referred to the notion of *normalisation*. Normalisation is defined by the Palestinian Boycott Divestments and Sanctions (BDS) Movement as

> [participation] in any project or initiative or activity, local or international, specifically designed for gathering (either directly or indirectly) Palestinians

(and/or Arabs) and Israelis, whether individuals or institutions, that does not explicitly aim to expose and resist the occupation and all forms of discrimination and oppression against the Palestinian people. (Palestinian BDS conference, Ramallah 2007, quoted in Kassis 2015)

Since 2011, the Palestinian liberation movement and its allies have progressively adopted anti-normalisation as a tactic, deliberately avoiding cooperation with institutions that do not explicitly oppose the Israeli Occupation. Both Amany and Micha framed their individual choices through the lens of this larger shift.

The encounter with the Israeli mappers highlighted the 'normalisation risks' associated with participation in OSM, prompting GJ to reflect on how their anti-normalisation stance should be applied to mapping. As the organisation became more familiar with OSM and its deliberative mechanisms the team discussed how to use the mapping platform strategically. Since I began working with GJ in 2015, the organisation has never engaged with other OSM mappers and has also largely refrained from contributing data to the project. Instead, they download OSM data and use it to produce their own maps, retaining full control over features and labels. This is more difficult to achieve for the interactive tile-based maps that populate the Web. Thus, although it is possible to set up an independent server to render the OSM map in Arabic, this has so far been out of GJ's technical reach. For this reason, the main map on the organisation's website contains a disclaimer that clarifies these choices:

Our homepage and PDF maps reflect our research independent of the borders drawn by the Israeli occupation government. The base layers of the interactive maps are from OpenStreetMap. GJ cannot control and does not endorse their information or terminology. (grassrootsalquds.net, August 2020)

GJ is a small organisation with strong political commitments: many Palestinian Jerusalemites will not share their views on the matter. Still, their choice not to map must be read as one instance of a broader strategy of non-engagement. In this case, non-participation should be considered as a form of political expression, rather than as a problem to be fixed. One may counter that the case of Jerusalem is exceptional, as is the 'local' notion of normalisation. Yet, tellingly, other researchers (Roth 2009; Wainwright and Bryan 2009) have suggested that mapping may not always benefit indigenous groups, especially when the 'rules' do not take into account their ways of knowing, needs and demands. Thus, similar non-participation dynamics are likely to be at play in other places and in other crowdsourced platforms. To recognise and investigate phenomena like political 'non-mapping' and hidden mapping controversies, it is essential to look not only beyond the rendered map, as many geoweb researchers have recognised (Glasze and Perkins 2015; Lin 2011), but also beyond crowdsourcing user communities, code and databases.

NAMES ON THE GROUND: A HISTORICAL DETOUR

The OSM map of Jerusalem comprises several hundred thousand nodes, and a similar amount of lines and polygons. Thus, regardless of the disputed note's symbolic importance, one could be forgiven for dismissing the controversy as an unfortunate occurrence. To fully grasp its significance, it helps to consider how the episode relates to the OSM mapping rules, in particular, the project's naming conventions. The OSM guidelines explain that the main name tags should contain names 'that are locally used, especially ones typically signposted', following the project's 'on the ground rule' (OSM Wiki 2018a). The 'on the ground rule' states that, when in doubt, users should 'map the world as it can be observed by someone physically there' (OSM Wiki 2018b). According to many OSM experts, this rule is what allows OSM to minimise editing disputes of the kind that afflict Wikipedia: 'when a conflict arises, the person who visits the location is going to be given deference over any other source' (Wroclawski 2014). The idea here is that site visits can settle disputes about the position and qualities of geographical objects in ways that are not possible for contentious encyclopaedic entries such as 'Islam' or 'Anarchism' (or indeed the spelling of 'yoghurt'). The expression 'typically signposted' indicates that street signs are considered the 'on the ground' evidence of a place's name. This principle is reiterated in the other Wiki pages that address the handling of disputes:

> When users do disagree on what should be included, or how a certain area should be represented, they are encouraged to work out a solution through deliberation. When this is not possible, the dispute should be settled by referring to what can be seen 'on the ground'. So, for example, in case of disagreement about a street name, the name that appears locally on the street sign should prevail. (OSM Wiki 2015)

And, explaining the rational for this decision:

> Sometimes there's conflicting information about, say, the name of a place. An old map might call it one thing, current maps another, and the place name sign something else. People using our maps (for navigation) won't care about the spelling in other maps, they need to find the names from local signs in the map and vice versa. (OSM Wiki 2016)

If users cannot agree on a main name tag, they can request the mediation of the OSM DWG.

In October 2011, one of the Israeli mappers involved in these discussions posted a message in the international OSM mailing list, hoping to receive support and advice from the wider OSM community.[2] This raised the controversy's profile and widened its scope. For the broader OSM community of users, the problem was not only finding a compromise about the Jerusalem node, but drawing out the dispute's implications for the project. Many viewed it as a symptom of a larger issue with the OSM rendering mechanism. The problem, several users agreed, is that by selecting only one of many alternative tags, the renderer leaves a wealth of data 'hidden' in the database. Two possible solutions were put forward. Some users suggested dropping the map altogether, thereby making clear that OSM is 'just' a data provider. Since the database can accommodate multiple tags describing the same property, such as name:he and name:ar, OSM would no longer be forced to choose on behalf of its users, and could therefore not be held responsible. On the other hand, some wished OSM would go in the opposite direction, offering multiple rendered maps based on different tags:

> OSM will never be able to create a *factual* map that is acceptable to all the world. (...) My suggestion of creating different maps, would do justice to all these differences, and while not creating the simple view of the world that many people want, may point, demonstrate and support the fact how we are all different in our view on this globe.

That, countered another user, would open up mapping wars everywhere, making the map useless: OSM is about facts, and it should not provide

[2] The quotes by international OSM mappers in this section are from the OSM-Talk forum threads '[OSM-talk] Naming dispute over Jerusalem - OSM failure' (October 2011) and 'Jerusalem name tag – Mediation' (October 2011).

'everybody with their comfy "virtual" world view'. Here, then, we see two conceptions of OSM. One emphasises usability, positioning OSM as a cost-free repository that allows the development of competitive map-based applications. The other stresses the visionary dimension of the project: an open map where anyone can contribute on an equal basis (see also Perkins 2014).

While no one disputed Jerusalem's status as an Israeli city and national capital in the local forum, international mappers wished to settle the matter by establishing globally valid criteria. They sought to resolve the question of Jerusalem's local language by agreeing on general parameters to define what counts as a 'local language'. So, for example, one user noted that Hebrew and Arabic were Israel's official local languages following a ruling by the Israeli Supreme court.[3] He also reflected that, if the local language is meant to serve local users, then what should count is the population make up. Another countered that population figures in the region were too contentious to be used for mediation. Yet another evoked the OSM 'on the ground' principle, and its emphasis on signage. Should not the map reflect the combination of Hebrew, Arabic and English that one encounters on many of the city streets?

But the language of street signs in Israel/Palestine is far from straightforward, since the region's linguistic landscape is fragmented and rapidly changing. A 2006 study (Ben-Rafael et al. 2006) found that the language of signs in Israeli public space follows different patterns depending on the local demographics. So, signs tend to be in Hebrew and English in areas with a Jewish majority, in Arabic and Hebrew in Palestinian centres within the 1948 borders, and in Arabic and English in East Jerusalem. Of course, Israel is not unique in this respect. Many countries where street signs are typically in more than one language have developed their own localised mapping standards on OSM. For example, users in India opted for English as their default language, while in Belgium they use both Dutch and French.

Thanks to the 'on the ground' rule, street signs become material proofs of name tag's objectivity and, by extension, of a place's linguistic and national identity. In this sense, the 'on the ground' rule adapts the epistemological principles of cartographic science to the populist

---

[3] This changed in July 2018, when Arabic was downgraded from an official language to one with 'special status'.

ethos of crowdsourcing. It takes the reality outside the map as a guarantee of truthfulness and invites the (imaginary) community of mappers to act as witnesses. Just like the witnesses of modern scientific experiments, mappers supposedly inhabit the 'culture of no culture' (Haraway 1997, Chapter 1; Shapin and Schaffer 1985): a generic crowd with no gender and no race, no political affiliation or subjective views. As I have just discussed, however, not everyone qualifies as a real mapper. In the remaining part of the chapter, I turn my attention to the street signs that supposedly guarantee the objectivity of the map. Their history, I argue, shows that the production of objects is a process as power loaded as the definition of suitable witnesses. A lengthy historical digression may seem out of place in a study of the geoweb, but it is necessary to fully grasp the political significance of the OSM ground rule in Jerusalem.

Under the Ottoman rule, street names fell beyond the state's realm of interest and control. A study published in Arabic in the 1920s, and reported by Kevin McCarthy (1975, p. 81), tellingly observes: 'Turkey has been, in all its projects, careless and indifferent as to most of its cities. It let the commoners name their neighbourhoods and streets as they wished'. Pre-Mandate maps of Jerusalem corroborate this observation. For instance, Fig. 5.1 shows portions of an 1883 map, where a combination of numbers, Latin and Greek letters mark the Old City's streets.

Starting from the 1900s, more systematic street names appeared in the newly established Zionist settlements, where they were used as markers of Jewish and Hebrew identity. This was achieved through the use of Hebrew language, as well as the celebration of Biblical figures and heroes of the Zionist movements (Azaryahu 1996; Bar-Gal 1989). By contrast, in Arab cities like Haifa, Jaffa or Jerusalem, Palestinian elites showed little interest in street naming, using vernacular names that typically referred to local topographical features, such as hills, sources of water or notable trees (Azaryahu and Kook 2002; Benvenisti 2002).

When the British established their mandate in 1914, they promoted street naming as a way to facilitate the administration of the newly acquired territories. The chosen names were meant to celebrate the cultural heritages of the city's different groups while minimising nationalist undertones (Azaryahu and Kook 2002). The result was a curious mix of Christian crusaders (e.g. Godfrey de Bouillon Street) and their Muslim adversaries (Suleiman), British royals (King George V Avenue), biblical kings (King Solomon Street) and Ottoman sultans (Selim I Road). All the

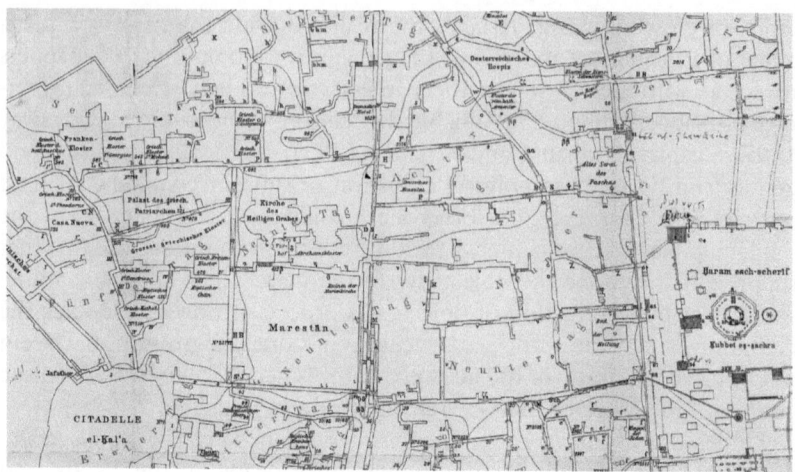

Fig. 5.1 Detail from Sandreczki. 1883. 'Plan Der Strassen & Plätze Des Jetzigen Jerusalem'. Zeitschrift des Deutschen Paelestina Vereins, Vol. VI [1883]. Related text: pp. 43–77. Public domain

same, street naming was from the start a source of political tension. Significantly, one of the founders of the Pro-Jerusalem Society, the organisation responsible for the task, noted:

> The names had to be in the three official languages, and the three traditions, Christian, Moslem and Jewish, had, so far as possible, to be preferred (...) Here was scope not only for scholarship but acute political division, and the sub-committee had on several occasions to be steered over very dangerous rocks. (Ashbee 1924, p. 26)

We can also credit the Pro-Jerusalem Society for setting the first street signs in the form of blue ceramic plaques, a characteristic design that still populates Jerusalem's historical neighbourhoods. They reported street names in English, Arabic and Hebrew, in this order, reflecting the official languages and their hierarchy. Names were translated in each language, rather than transliterated. So, for example, 'Street of Prophets' was inscribed as 'Tariq Al-Anbiya' in Arabic and 'Rehov Ha-Nevi'im' in Hebrew (see Fig. 5.2).

Fig. 5.2   British Mandate street sign. The name 'Street of Prophets' is translated in three languages, English, Arabic and Hebrew (*Source* Photographed by user DMY in 2007, Wikipedia Commons [licensed under CC BY 3.0])

In 1938, the municipality appointed a special Street Naming Committee (SNC) to prepare a set of guidelines for Jerusalem street names. The committee, which included Christian, Jewish and Muslim representatives, divided the city into a predominantly Jewish area, a predominantly Arab area, and a mixed area. Jewish members of the SNC proposed names for the Jewish zone, Arab members for the Arab zone, and the whole committee was involved in the naming of the mixed areas.

With the 1949 partition of the city, the West Jerusalem administration prioritised changing street names and signs so as to reflect the newly established Jewish sovereignty (Azaryahu 2012). The new signs dropped English names and placed Hebrew names first. There was also a shift from translation to transliteration: Street of Prophets became 'Ha-Neviim' in English. English-language scholarship has not studied the Jordanian administration's policies to the same degree as those of West Jerusalem. Nonetheless, signs from this era, like the one in Fig. 5.3, indicate that naming practices were reversed in East Jerusalem, with Hebrew dropped, and Arabic placed first.

Fig. 5.3 A street sign in the Old City's Muslim Quarter. The sign illustrates the struggle for linguistic dominance over Jerusalem. Its bottom portion, put up after the Jordanian administration came into power, gives the Arabic name above the English transliteration. The Hebrew name was added above the other names after 1967 (*Source* Photographed by user Yoav Dothan in 2009, Wikimedia Commons, Public Domain)

Maps produced between 1949 and 1967 by either Jordan or Israel tend to both blank out 'the other side', representing it as empty or full of rubble. Furthermore, many street names were changed in this period to match the city's new national identities. On Israeli maps, one can see that St Paul's Street, in the West of the city, has become Shivtel Israel, i.e. Tribes of Israel; from Jordan maps, we learn that Suleiman Street had been temporarily renamed King Hussein Street. However, Jordanian authorities seemed uninterested in extending their street-naming system beyond the Old City and the adjacent commercial areas. Officers of the Israeli administration report that most streets in the east remained unnamed when Israel took over in 1967.

The first systematic efforts to fill these gaps started in the 1980s, after the Jerusalem Law sanctioned the city's status as the United Capital of Israel (see Chapter 1). One episode, reported in the book *Separate and Unequal* (Cheshin et al. 1999, pp. 146–148), powerfully illustrates the kind of political arm wresting involved in the process. When Israeli authorities decided to name the streets of Beit Hanina, a Palestinian neighbourhood north of Shu'fat, they asked the local Palestinian leader to consult with residents and provide a list of potential names for submission to the municipal Name Committee:

> It was several weeks before the list was sent to city hall. As municipal officials reviewed the list, they slowly began to understand its significance. All the names were of Arab villages that had existed before 1948 but were destroyed by Israel during the war: Umm Rashrash, Banias, Majdal, Askalan, Yaffa, Pluga, and others. The municipality contacted Darwish, and he unabashedly explained the neighborhood council's idea: "We see the map of Beit Hanina as representing that of all Palestine," Darwish said. "In the north of Beit Hanina, we will give the streets names of the villages that once stood in northern Palestine, in the west of the neighborhood, the roads will have the names of the villages that once stood in the west of Palestine, and so on."

Darwish was told to try again with something less political, such as flowers and trees: the municipality would not accept such an expression of Palestinian nationalism on its streets. Darwish followed the instructions to the letter, and submitted a list of flowers and trees, which was promptly approved by the Name Committee. However, the municipality delayed the posting of the chosen names for years, with justifications like:

> "We can't send a work crew to put up the sign because it is too dangerous in the neighbourhood"; "We can't go to east Jerusalem during the intifada"; "The border police unit that was supposed to accompany us cancelled at the last minute."

When Cheshin and colleagues published their book in 1999, Beit Hanina streets still had no signs, although that has changed in the years since. Nearly twenty years after the neighbourhood's annexation, the municipality still lacked the sovereignty necessary to name streets.

The continued importance of street signs as political signifiers is attested to by the frequent proposals to standardise the country's toponomy by adopting the Hebrew names and transliterating them in the Latin and Arabic alphabets: Jerusalem would then become Yerushalaim in both Arabic and English. Commenting on one such proposal, Transport Minister Yisrael Katz was quoted as saying:

Some Palestinian maps still refer to the Israeli cities by their pre-1948 names. I will not allow that on our signs. This government, and certainly this minister, will not allow anyone to turn Jewish Jerusalem to Palestinian al-Quds. (BBC News 2009)

Even if the proposed policy was not approved, it has become common for signs to translate the Hebrew *Yerushalaim* in Arabic, rather than using the more common Arabic name Al-Quds.

The municipality publishes a list of existing street names on its website. It contains 3285 streets, ordered by neighbourhood. Of these, 2920 (88%) also appear in the OSM database.[4] The list suggests that streets in some Palestinian neighbourhood were still largely unnamed as late as 2014. For example, the neighbourhood of Kufr' Aqab, formally part of the municipality but cut off from the city by the West Bank Barrier, does not even appear on the list. Only four names are associated with Ras al-Amud. In the case of Um Tuba, the list provides twenty-nine names, yet only two appear on OSM. This suggests that few OSM users are mapping the area, but could also mean that the names reported in the list have not yet been posted and thus exist, for the time being, only in the municipal records.

For many, the absence of street names in Palestinian neighbourhoods demonstrates the municipality's neglect of these areas, since it makes delivering basic services almost impossible, including a door-to-door postal system and the provision of fire protection and ambulances (Cheshin et al. 1999). Others (Coffey 2014) highlight the effects of these gaps on a discursive level: by presenting these localities as empty, they effectively erase their Palestinian inhabitants. When I first started working on this topic, I also shared this view (Carraro 2015). Here, however, I have argued for a different interpretation: blank areas on the map mark gaps in the municipality's authority. As the Israeli municipality has tightened control over Palestinian neighbourhoods like Shu'fat and Beit Hanina, it has named their streets and appointed them with signs. This slow process remains incomplete: in many areas, names either do not exist or are not posted on site. Without marked street names, taxes

---

[4] The comparison between OSM tags and the names on the municipality's list was done using SQL queries in ArcGIS. This method has a relatively large probability of error, since any difference in the spelling of the two names will result in a missed match: for example, 'Ras al Amud' and 'Ras al-Amud' will be read as different names.

and fines are more easily avoided. Strangers, including police and military, hesitate to venture into unfamiliar, hostile neighbourhoods. Visitors looking at the city on a map may intuit that those areas do not quite belong to the 'normal', reunited Jerusalem.

* * *

When the Jerusalem node controversy took place, the city was going through a street-naming spree, focusing on East Jerusalem: in 2012, alone the municipality issued 145 new street names (Associated Press 2012), and its efforts have continued in the years since. Predictably, name choices have been loaded with political tension. For example, in 2015, 30 streets in the central Palestinian neighbourhoods of Silwan, Sheikh Jarrah and Mount of Olives were given Hebrew names, in spite of protests against what was denounced as an attempt at 'Hebraising' the city (Nazzal 2015). By then, the municipality could claim to have named, mapped and sign-posted almost 90% of East Jerusalem streets (Miller 2015). In one of our conversations about these developments, a GJ staff member downplayed these concerns, noting that Palestinians should not be fighting for Arabic names, but for the liberation of Jerusalem. The problem, she insisted, lies in the Israeli municipality's power to name the city's streets, rather than in its questionable choices.

To those familiar with the language of international diplomacy, the 'on the ground' rule may well bring to mind the phrase 'facts on the ground', which refers to 'changes in human geography engineered by a state to strengthen a political claim' (Berridge and Lloyd 2012, p. 147). This expression is often used to discuss Israeli policies in the West Bank and East Jerusalem, particularly in reference to the expansion of settlements (e.g. Hodgkins 1998; Khalidi 2010). As Berridge and Lloyd note (2012), this strategy is most frequently employed by states that have the means and influence to make these changes, but do not have full sovereignty. It is useful to think of these two expressions in conjunction because, while nearly identical, they emphasise different properties of ground objects. The OSM principle emphasises that material objects, say, street signs, cannot be ignored. As the OSM editing guidelines note, if people are to use the map for navigation, its labels must match those on local signs (OSM Wiki 2016). The language of diplomacy, on the other hand, reminds us that what is on the ground has been laboriously constructed, often with specific political goals.

The OSM database allows users to create a virtually endless number of alternative name tags to be incorporated in custom-based map renderings, generating a sort of 'cartographic pluralism'. These, however, need to be stored under the appropriate tags: alternative 'points of view' separated from the 'objective data' that is part of the main map. The main data are widely circulated 'by default' through countless apps and websites. Alternative points of view can be used, but at greater cost, requiring higher technical skills and a private server. This separation ensures OSM's usability, offering developers, cartographers, GIS analysts and researchers a database that competes in many respects with proprietary ones, but the exclusion of alternative tags is difficult to reconcile with ambiguity, indeterminacy and open disagreement.

REFERENCES

Ashbee, C. R. (1924). *Jerusalem, 1920–1922, Being the Records of the Pro-Jerusalem Council During the First Two Years of the Civil Administration.* London, Murray. http://archive.org/details/jerusalem192019200ashbuoft. Accessed 13 July 2018.

Associated Press. (2012, November 1). Where the Streets Have No Names No More: East Jerusalem Gets Street Signs. *Associated Press.* http://www.foxnews.com/world/2012/11/01/where-streets-have-no-names-no-more-east-jerusalem-gets-street-signs.html. Accessed 23 May 2017.

Azaryahu, M. (1996). The Power of Commemorative Street Names. *Environment and Planning D: Society and Space, 14*(3), 311–330.

Azaryahu, M. (2012). Hebrew, Arabic, English: The Politics of Multilingual Street Signs in Israeli Cities. *Social & Cultural Geography, 13*(5), 461–479. https://doi.org/10.1080/14649365.2012.698748.

Azaryahu, M., & Kook, R. (2002). Mapping the Nation: Street Names and Arab-Palestinian Identity: Three Case Studies. *Nations and nationalism, 8*(2), 195–213.

Bar-Gal, Y. (1989). Cultural-Geographical Aspects of Street Names in the Towns of Israel. *Names, 37*(4), 329–344. https://doi.org/10.1179/nam.1989.37.4.329.

BBC News. (2009, July 13). *Row Over 'Standard' Hebrew Signs.* http://news.bbc.co.uk/2/hi/8148089.stm. Accessed 4 July 2018.

Ben-Rafael, E., Shohamy, E., Amara, M. H., & Trumper-Hecht, N. (2006). Linguistic Landscape as Symbolic Construction of the Public Space: The Case of Israel. *International Journal of Multilingualism, 3*(1), 7–30. https://doi.org/10.1080/14790710608668383.

Benvenisti, M. (2002). *Sacred Landscape: The Buried History of the Holy Land Since 1948*. Berkeley, CA and London: University of California Press.

Berridge, G., & Lloyd, L. (2012). Facts on the Ground. In *The Palgrave Macmillan Dictionary of Diplomacy*. Basingstoke: Palgrave Macmillan.

Bittner, C. (2014). Reproduktion Sozialräumlicher Differenzierungen in OpenStreetMap: das Beispiel Jerusalems. Reproduction of Social Fragmentations in OpenStreetMap: the Example of Jerusalem. *KN—Journal of Cartography and Geographic Information, 64*(3), 136–144. https://doi.org/10.1007/BF0354 4143.

Bittner, C. (2016). OpenStreetMap in Israel and Palestine—'Game Changer' or Reproducer of Contested Cartographies? *Political Geography*. http://www.sciencedirect.com/science/article/pii/S096262981630035X. Accessed 27 December 2016.

Carraro, V. (2015, March). *Experiences from Occupied Jerusalem*. Presented at the Doreen Massey Annual Event, Open University, Milton Keynes. http://www.grassrootsalquds.net/sites/default/files/20150506_Valentina_Carraro_DM7_OU_2015_copy_0.pdf. Accessed 16 July 2018.

Cheshin, A., Hutman, B., & Melamed, A. (1999). *Separate and Unequal: The Inside Story of Israeli Rule in East Jerusalem*. Cambridge, MA: Harvard University Press. http://public.eblib.com/choice/publicfullrecord.aspx?p=3299990. Accessed 22 May 2017.

Coffey, Q. (2014, July 14). Google Maps in Palestine. *openDemocracy*. http://www.opendemocracy.net/north-africa-west-asia/quinn-coffey/google-maps-in-palestine. Accessed 23 May 2017.

Glasze, G., & Perkins, C. (2015). Social and Political Dimensions of the OpenStreetMap Project: Towards a Critical Geographical Research Agenda. In J. J. Arsanjani, A. Zipf, P. Mooney, & M. Helbich (Eds.), *OpenStreetMap in GIScience* (pp. 143–166). Springer International Publishing. https://doi.org/10.1007/978-3-319-14280-7_8.

Haklay, M. and Budhathoki, N. R. (2010). OpenStreetMap—Overview and Motivational Factors. Powerpoint Presentation. Horizon Infrastructure Challenge Theme Day, Nottingham. https://www.ideals.illinois.edu/bitstream/handle/2142/16461/Horizon%20March%202010%20(Haklay%20and%20B udhahtoki).pdf. Accessed 31 August 2021.

Haraway, D. (1997). *Modest_WitnessSecond_Millennium. Female-Man©_Meets_OncoMouseTM: Feminism and Technoscience*. New York and London: Routledge.

Hodgkins, A. B. (1998). *Israeli Settlement Policy in Jerusalem: Facts on the Ground*. Jerusalem: Passia.

Kassis, R. O. (2015, June 3). Thoughts on Normalisation in the Israel-Palestine Conflict. *No2brandisrael*. http://www.no2brandisrael.org/thoughts-on-normalisation-in-the-israel-palestine-conflict/. Accessed 27 September 2017.

Khalidi, R. (2010). *Palestinian Identity: The Construction of Modern National Consciousness*. New York: Columbia University Press.

Lin, Y.-W. (2011). A Qualitative Enquiry into OpenStreetMap Making. *New Review of Hypermedia and Multimedia, 17*(1), 53–71. https://doi.org/10.1080/13614568.2011.552647.

McCarthy, K. M. (1975). Street Names in Beirut, Lebanon. *Names, 23*(2), 74–88.

Miller, E. (2015, September 21). 30 East Jerusalem Streets given Hebrew Names, Enraging Arab residents. *Times of Israel*, online.

Mr_Israel. (2011, August 1). 'Edit War' over Jerusalem—Meeting with East Jerusalem Mappers. *OpenStreetMap Forum/Israel*. https://forum.openstreetmap.org/viewtopic.php?id=13178. Accessed 6 January 2017.

Naughton, J. (2019, September). In a Hysterical World, Wikipedia is a Ray of Light—and that's the Truth. *The Guardian*, online.

Nazzal, N. (2015, September 19). Jerusalem Street Names to Be Hebraised. *GulfNews*. http://gulfnews.com/news/mena/palestine/jerusalem-street-names-to-be-hebraised-1.1589133. Accessed 22 May 2017.

OSM Wiki. (2015, July 8). Data Working Group/Disputes. *OpenStreetMap Wiki*. http://wiki.openstreetmap.org/wiki/Data_working_group/Disputes#Jerusalem%2520Name%25. Accessed 3 January 2017.

OSM Wiki. (2016, June 28). Good Practice. http://wiki.openstreetmap.org/wiki/Good_practice#Map_what.27s_on_the_ground. Accessed 29 August 2016.

OSM Wiki. (2018a, June 20). Key:name. *OpenStreetMap Wiki*. https://wiki.openstreetmap.org/wiki/Key:name#cite_note-1. Accessed 6 July 2018.

OSM Wiki. (2018b, September 6). How We Map. http://wiki.openstreetmap.org/wiki/How_We_Map. Accessed 29 August 2016.

Perkins, C. (2014). Plotting Practices and Politics: (Im)Mutable Narratives in OpenStreetMap. *Transactions of the Institute of British Geographers, 39*(2), 304–317.

Roth, R. (2009). The Challenges of Mapping Complex Indigenous Spatiality: From Abstract Space to Dwelling Space. *cultural geographies, 16*(2), 207–227. https://doi.org/10.1177/1474474008101517.

Shapin, S., & Schaffer, S. (1985). *Leviathan and the Air-Pump: Hobbes, Boyle, and the Experimental Life*. Princeton, NJ: Princeton University Press.

Stephens, M. (2013). Gender and the GeoWeb: Divisions in the Production of User-Generated Cartographic Information. *GeoJournal, 78*(6) 981–996. https://doi.org/10.1007/s10708-013-9492-z.

UNDP. (2014). *The 2014 Human Development Report—Education* (p. 5). UNDP.   https://www.google.com/url?sa=t&rct=j&q=&esrc=s&source=web&cd=&cad=rja&uact=8&ved=2ahUKEwjVuZa8n5bsAhU_HbkGHc7FBbEQFjAAegQIBxAC&url=https%3A%2F%2Fwww.undp.org%2Fcontent%25

2Fdam%2Fpapp%2Fdocs%2FPublications%2FUNDP-papp-research-PHDR20
15Education.pdf&usg=AOvVaw05Ou52LUO4Yh4N8l4dtTCq. Accessed 2
October 2020.
Wainwright, J., & Bryan, J. (2009). Cartography, territory, Property: Postcolo-
nial Reflections on Indigenous Counter-Mapping in Nicaragua and Belize.
cultural geographies, *16*(2), 153–178. https://doi.org/10.1177/147447400
8101515.
Wroclawski, S. (2014, January 17). Edit Wars in OpenStreetMap. *Emac-
sen's Blog*. http://blog.emacsen.net/blog/2014/01/17/edit-wars-in-openst
reetmap/. Accessed 11 October 2016.

CHAPTER 6

Epilogue

Abstract This last chapter links together the three case studies, drawing some overall conclusions. The common theme that emerges from the three examples, I suggest, is that Palestinians in Jerusalem are often 'gaps on the map'. While it is certainly possible to read their absence as evidence of political exclusion and marginalisation, I argue that it may be more accurate to interpret it as an instance of 'resistance to being enrolled' into the Israeli municipality. This alternative framing underscores that a more complete and balanced representation of Palestinians on the geoweb is not helpful in the absence of broader changes to Jerusalem's political and spatial organisation.

Keywords Jerusalem · Palestine · Future · Critical cartography · Digital technologies

Geoweb applications are at once similar to and immensely different from traditional paper maps. They *look* different, they rely on new data sources processed through computer software, and they 'follow us' everywhere we go, offering many new functionalities that would seem science fiction-like to the surveyors who first charted Palestine in the nineteenth century. Yet, they are still intensely political: they represent and exclude, they format space and populations, legitimate and spread geographical 'facts', carry symbolic values and, often, provoke heated debates. The starting point of my research was that, to better understand how emerging technologies reconfigure these processes, there is a need to develop new approaches to the study of maps, without forsaking the sensibilities that critical cartographers have taught us to cultivate.

To this end, I have examined here three Jerusalem-related mapping controversies, exploring the debates they generated through user interviews, participant observation and the analysis of user forums, social media and news reports. While it can be useful to separate these disputes for the purpose of analysis, where we draw the boundaries is a political question. Examining them together allows us to see how they contribute to define each other's significance, 'adding up' to the broader 'political situation' (Barry 2012) investigated by this research: namely, the enactment of Jerusalem through web maps. More concretely, the three disputes point at a general pattern in the web cartographies of Jerusalem, namely the omission of Palestinians and their localities. This takes different forms, such as lack of labelling, patchy datasets, erratic routing algorithms and scarce participation in VGI platforms. In the case of Google Maps, the omission likely stemmed from a mistake, but served to highlight that Palestine is still a contested issue—a matter of care—for large publics. In the other two controversies, gaps in the map are directly informed by Jerusalem's political setting. More specifically, they point at places where state control is weak and Israeli authorities struggle with fundamental tasks such as guaranteeing the safety of their citizens or imposing a systematic toponomy. The common theme that emerges from these controversies is the precarity of the 'fact' of Palestine, not only at a discursive level—because its existence is a disputed matter—but also 'on the ground': Palestine exists, but only in a very partial and incomplete way. To conclude, I want to advance some possible interpretations of these omissions and draw out their implications for how we should go about researching, thinking of and engaging with, geoweb maps and other digital technologies.

Gaps in the Digital Map

Recent studies on the cartographic representation of Israel/Palestine on the geoweb have consistently noted that Palestinian views are marginalised (Bittner 2014, 2016; Ford and Graham 2015; Graham and Zook 2013). The examples examined in this book support these findings and extend them by considering other aspects of the map, beyond its informational content. The mapping providers under study all seem to favour an Israeli perspective on Jerusalem. In the case of Google Maps, this is achieved through the curation of data sources that support Israel's claims on the city; in Waze, by equating 'users' with 'Israeli Jews', and 'danger' with 'Palestinians'; and on OSM, Israeli perspectives are favoured by low participation rates among Palestinians and by the privileging of 'on the ground' verification.

In this respect, then, my research also speaks to current debates about the emancipatory potential of the internet beyond the cartographic literature. Especially in their early years, internet platforms, and especially social media sites, were seen as tools through which people could confront and even overthrow oppressive political regimes: they (supposedly) gave voice to ordinary citizens, allowing them to sidestep censorship and galvanise grassroots mass movements, as in the case of the 2009 protests in Iran or the Arab Spring in 2010–2012. Even if optimism about Internet media has considerably waned in the intervening decade, hope in a technological solution to social problems is surprisingly tenacious (the many apps promising to 'optimise' your finances, health and stress levels being a small, everyday manifestation of that hope). My analysis underscores the dangers associated with overstating the transformative power of online discussion spaces like forum threads, VGI platforms, social and news media. Inevitably, these are embedded in their offline settings, shaped by the uneven visibility of Israeli and Palestinian sources, the profit-making logic of commercial platforms and prevalent geopolitical imaginaries. These constraints are evident in each of the disputes I examined. In the Google Maps controversy, discussions surrounding Palestine's erasure tended to be superficial, misinformed and short-lived. In the second case study, Palestinians were strikingly silent, despite being one of the groups most likely to be affected by WADA. Finally, despite its theoretically open structure, OSM seemed dominated by Israeli expert mappers, as evidenced, for example, by the tags' language patterns and the discussions on the project forum.

Technologies such as those discussed here (map mashups, the semantic web, navigational algorithms, crowdsourcing, online forums and social media) must be positioned into the wider socio-economic and cultural context in which they operate. For many critical geographers (e.g. Lee 2010; Sadowski and Pasquale 2015; Shaw and Graham 2017), this means analysing the geoweb through the lens of political economy, interpreting the emergence of new mapping technologies as emblematic of 'a shifting relationship between the state and market entities such as corporations in accordance with neoliberal rationalities that are uniquely structuring the production and dissemination of geographic information' (Leszczynski 2012, p. 77). On this account, digital media, including digital maps, are the product of broader social relations and are therefore likely to reflect and reproduce those relations, reinforcing socio-economic and political inequalities between groups. The case of Israel/Palestine is often seen as emblematic of this general trend (Bittner 2016; Leuenberger and Schnell 2010; Wood 2010a).

The controversies around the maps of Jerusalem, however, demonstrate the importance of local settings in shaping how these technologies work. While the geoweb features under study are at work in many parts of the world, they are transformed through their usage in the Jerusalem context. In other words, web maps are informed not only by their code, usage policies and profit models (elements that tend to remain similar, though not identical, everywhere), but also by 'local' actors, including a place's history, current social and political problems, spatial patterns and user cultures. For example, Google Maps is generally seen as advancing hegemonic discourses through its cartographic representations, closing off the possibilities for dissent through its 'black-boxed' algorithms (Fuchs 2011; Shaw and Graham 2017; Zook and Graham 2007a). However, in the case examined here, a technical glitch worked as a device for political engagement, leading a large international public to question how maps routinely (fail to) represent Palestine. Safety navigation apps like Waze are seen as leading to greater securitisation and spatial segregation by drawing on existing ethnic and racial stereotypes (Amoore 2009; Leszczynski 2016; Thatcher 2013). This interpretation seems inadequate to the context of Israel/Palestine, where a complex bordering apparatus already divides Israelis and Palestinians, constituting them as different political subjects. Instead, I have suggested that Waze's datasets and algorithms highlight the discrepancies between competing definitions of the Israeli nation-state. Finally, most studies understand under-representation on crowdsourcing platforms as the result of scarce participation because of economic and social marginalisation (Ford and Wajcman 2017; Graham et al. 2014). In the case of Jerusalem, however, other factors are also at play, including a deliberate anti-normalisation strategy on the part of Palestinians and the spatial distribution of place names and signs beyond the map.

A corollary of these observations is that the effects of digital mapping technologies are more ambiguous than is generally assumed. Granted, they reflect particular ways of knowing, supporting political arrangements and knowledge claims that often favour powerful actors, be it Google or the Israeli state. However, they are not invincible, nor do they work towards a coherent project of social control, carried out by invisible hands, as at times the critical cartography literature implies (on this point see also Rose-Redwood 2015). Cartographic technologies are not indifferent to the 'real' city that extends beyond the map: the positions where rival armies stopped fighting, the street signs, the concrete barriers, the railway lines and bus stops, the languages people speak, the areas where police forces cannot reach, where residents rebel or compromise. There is no overarching order guiding the process, leaving room for unforeseen results.

Paradoxically, in the cases I present here, through bugs, incidents and debates, web maps have also generated opportunities for calling into question the 'reality' of Israel as a Jewish democratic state and highlighted the presence of Palestinians in Jerusalem and beyond. In the case of Google Maps, the service's visibility has created a space for Palestinians to enact a shared national identity online, and for their supporters to voice solidarity. Waze marks the presence of Palestinians within areas that are considered non-negotiably Israeli, revealing the inconsistency between the State's spatial and ethnic definitions. Meanwhile, OSM's uneven data coverage shows that the municipality's 'on the ground' control in many Palestinian areas is still far from complete. In short, the translation of an uncertain matter, such as Palestine's existence, into cartographic form is a process susceptible to stoppages, faults and failures. In these instances, web maps draw attention to the instability of 'facts', rather than affirming their certainty. From this point of view, the near-absence of Palestinians from online maps could be recast as a sign of resistance to being enrolled (Star 1990), i.e. to integration into the Israeli municipality. In many parts of Jerusalem, Israeli sovereignty is incomplete, as evidenced, for example, by the sense of danger (real or perceived) experienced by Jewish Israelis driving in certain neighbourhoods or by the inconsistent signposting of street names. When we pay attention to these nuances, it becomes clear that geoweb maps do not record spatial patterns in a neutral way, but nor are they all-powerful tools for projecting social structures onto an empty canvas. Rather, they are hybrid objects—neither 'natural' nor 'social'. The

'power architectures' that structure the working of geoweb technologies are susceptible to failure and change, generating openings that allow
unincluded actors to make their voices heard.

This interpretation also draws attention to the fact that Palestinians
are not passive victims of a map, but mapping actors: they decide not to
map, they resist being mapped, they use maps in political ways without
directly participating in their making. Through these remarks, I do not
mean to deny or minimise the disadvantages that certainly stem from
cartographic under-representation, both materially, for example in terms
of economic losses for businesses, and symbolically. However, one must
wonder whether edits to the map would be meaningful in the absence
of changes to the city's spatial and political organisation. This way of
thinking about the gaps in the Jerusalem map challenges the prevalent
liberal understanding that assumes cartographic controversies ultimately
revolve around questions of representation. In this view, when people
disagree about a map, it is because they hold different views of the same
reality; if all perspectives are fairly represented, through dialogue, it should
be possible to resolve these differences and reach consensus. Yet, as the
case of Jerusalem so effectively demonstrates, cartographic conflicts often
extend beyond the map. Peace will not be achieved by all parties agreeing
on the city's 'true' nature, but by their ability to build a version of
Jerusalem most can live with and in (see also Latour 2004, p. 455 for
more on this point).

In sum, paying attention to the specificities of mapping in Jerusalem
reveals the distinctive interplay between the geoweb and the Israeli settler
colonial project. Geoweb maps contribute to assemble the 'reality' of
Jerusalem, which also includes the making of Palestinians as second-
class people: non-citizen, non-Jewish, non-white, at once dangerous and
non-existent. The cartographic erasure of Palestinians goes hand in hand
with the territorial expansion of the Israeli state: thus, the gaps in the
maps of Jerusalem are best understood as the latest iteration of long-
standing narratives about Palestinian non-existence, rather than as the
product of socio-economic disadvantage or political under-representation.
Importantly, given the eurocentrism of geoweb research, it is difficult to
tell whether Jerusalem is an exception, or whether existing interpretations need to be revised. To find out, we need to expand the range of
empirical case studies in geoweb research. What is clear, however, is that
geoweb studies need to pay greater attention to the transformations that

technological objects undergo as they become embedded in particular places.

Admittedly, my emphasis on instances when web maps have worked to catalyse dissent rather than supporting consolidated 'facts' about Jerusalem stems from my research approach, which has taken carto-graphic controversies as its starting point. In other words, I have focused here on 'visible' problems that had been extensively discussed by promi-nent actors, such as mainstream media, politicians and experienced OSM mappers. Google Maps, Waze and OSM, however, operate mostly 'in the background', serving billions of users every day. In most cases, their effects go unnoticed, but this of course does not mean they do not matter. Often it takes a dramatic event—a Twitter storm, a shooting, an army intervention—to gather a public. While it is certainly not my intention to advocate for research that focuses on sensationalistic case studies, I believe that controversy analysis can help to establish connections between visible disputes and broader issues that are out of the public eye. I believe this approach can add complexity to current discussions about the geoweb. This is because there is a persistent tendency—both inside and outside academia (e.g. Amoore 2009; Crampton 2003; Sadowski and Pasquale 2015)—to pit (mapping) technologies against people: the first advancing a terrifying project of social control, the latter struggling for freedom. The cases presented here challenge such schematic, black-and-white narratives. Like human actors, mapping objects—datasets, measurements, algorithms and graphic interfaces—are neither 'good' nor 'evil', though they do 'good' and 'evil' things. Their effects are varied and always changing: the power of maps, it turns out, can also work to generate and publi-cise disagreement. This way of engaging with maps may offer scholars and activists an alternative to the deconstructive analyses that make up the bulk of the critical cartography literature. These contributions call maps into question: they expose the commodifying logic of Google Maps (Zook and Graham 2007b), reveal the racist discourses that underpin safety navigation apps (Leszczynski 2016), or deconstruct the episte-mology of crowdsourced projects (Sieber and Haklay 2015). They are very effective at challenging online representations, but risk leaving the 'real' world untouched. By contrast, my account does not claim that the map is 'wrong', recognising that it is shaped, among other things, by the objective reality of battle lines, on the ground control and street signs. This relation is not taken as a 'realist' guarantee of the map's exactness, however, but rather as a starting point for challenging what happens beyond the map.

No Escape in Technology

I have focused here on three controversies related to mainstream maps. My intention was to explore the issues that surround popular mapping providers, which produce the sort of maps that are likely to reach a large number of people, and inform the views and actions of visitors and internet users without a specific interest in Israel/Palestine or Jerusalem. I have not discussed any of the many counter-mapping projects that geoweb technologies have helped to create. Perhaps the most famous case is Palestine Remembered (palestineremembered.com), briefly mentioned in Chapter 4 and examined in depth by Linda Quiquivix (2014). Alongside an archive of photos and historical texts, the project includes a comprehensive map of Palestinian villages made in Google Earth. The villages are divided into four categories: those that were destroyed or depopulated, those that have been occupied since 1948, those that have been occupied since 1967, and those that have been replaced by Jewish settlements. Developed through the crowdsourcing of oral histories, the map exemplifies how online media can be appropriated to challenge dominant, state-supported narratives, such as the customary compartmentalisation of Palestinians into 'Arab Israelis', residents in the Palestinian Territories and refugees in the diaspora (Quiquivix 2014, p. 455). Another, more recent example is Palestine Open Maps (palopenmaps.org), launched in 2020, which uses 1940s British Mandate maps as a base layer, with the aim of combining them with historical statistics and photos, oral histories and present-day data. Both projects illustrate the weight of the past on contemporary Palestinian mapping practices, mapping practices that leading critical cartography practitioner and scholar Denis Wood (2010b, p. 246) has described as 'reactionary'. For Wood (2010b, p. 141), this kind of counter-mapping forces indigenous people to adopt the language of the colonisers to engage in a cartographic struggle that resembles 'schoolyard name-calling—"You map me, huh? I map you!!"' While this remark is not untrue, there is a risk of essentialising indigenous knowledge, portraying it as static and confined to tradition, while overlooking both the long tradition of Arab cartography (Pursley 2015) and the more recent efforts to rethink science through indigenous frameworks (Walter and Andersen 2013). As for the backwards looking tendencies of much Palestinian cartography, it is important to underscore that, in the absence of a state, Palestinians must rely on this kind of popular practice to record and document the history of the Nakba. Crucially, it is a history that is not actually past, but continues to determine the present and will need to be reckoned with to realise a better future.

In his short story 'N', Palestinian journalist and author Majd Kayyal imagines a world where Israelis and Palestinians do not have to share land and can build their respective states, from the river to the sea. It is 2048 and, thanks to Israeli technology, the land has been cloned. Israeli Jews have achieved what they wanted all along, 'an entire nation exclusively for their "pure blood"'. And Palestinians have finally been given back their land, or at least a faithful copy of it: almost perfect, except that the sea around it extends for only ten kilometres, depleting Palestinian fish stocks and making grouper 'rarer than the rarest gem'. Refugees have been able to return to their homes, be they in Gaza, Nablus, Haifa or Jaffa. People on both sides have had to choose where they want to live: the Palestinian reality, or the Israeli reality; only those born after the ratification of the peace agreement can freely travel from one to the other, in a likely reference to the relative freedom of movement enjoyed by Palestinians living in the 1948 territories. This future world is far from perfect: Islamist groups have sprung up, along with 'corrupt real estate companies specialised in refugee property'; meanwhile, many people have become addicted to virtual realities, a market controlled by American and Japanese companies intent on producing war and porn scenarios. Still, a century after the Nakbah, technology has created an equitable solution to the conflict, finally bringing peace.

As of 2020, the option to clone the land remains sci-fi material, but the story plays on the real hopes of techo-utopians everywhere: technology providing a fix to thorny political problems, removing the need to address the root of the conflict—to concede, reconcile, compensate or reconsider. In Kayyal's story, each party gets exactly what they wanted. There is, however, a catch: Article 7 of the peace agreement prohibits both sides from commemorating the hostilities, or from mourning or celebrating those past events. 'We accepted a life in the present, with no past', recognises one of the narrators, '(...) That is what turned our glorious victory into a defeat'. Like the players of those all too popular virtual realities, people must go through their new lives without their collective memory, anonymous and isolated. In these conditions, Israelis cannot free themselves from Zionism; Palestinians, with no way to heal from their historical trauma, cannot let go of their resentment. Kayyal's story reminds us of the limits to technology's power to bring about change. Technologies like those associated with the rise of the geoweb are powerful, for the better or the worse. At best, when they are designed with the interests of their users in mind and shared widely, they can help people navigate their surroundings, access new information, share their views or communicate with one another. They can make life a little bit easier, in small but tangible ways. We should not, however, expect them to offer purer, transcendent truths. Nor should we hope they can solve complex political problems, delivering the just, peaceful worlds that we have so far not managed to create for ourselves.

In the years I have spent working on this project, at conferences, seminars and gatherings with friends, I have often been asked variations of the question: do new mapping technologies make cartography overall more just or more oppressive? Can a feature like WADA be redesigned to keep *everyone* safer, or is its concept inherently problematic? Is Google Maps a better or worse gatekeeper compared to national ordinance surveys? Is OpenStreetMap really a 'dooacracy' (Glaze and Perkins 2015) dominated by tech-bros, or do OpenStreetMap's undeniable qualities outweigh its limitations? Faced with these questions, I am reminded of Ursula K. Le Guin's splendid essay (2017, p. 17) on technological utopias. For Le Guin, utopia (intended as a place where people can live more comfortably, more equitably, more freely) is neither behind nor ahead of us. The question puts us in 'a rational dilemma, an either/or situation as perceived by the binary computer mentality, but neither the either nor the or is a place where people can live'. Therefore, 'the only answer one can make, I think is: No'. It is misleading to characterise the transformations brought about by the web as a rupture with the past, whether positive or negative. Rather, web technologies reframe the possibilities for political representation and action. If the future of Palestine depends also on whose 'facts' will be allowed to circulate, online and offline, then it is important for users and researchers to develop a more nuanced understanding of the opportunities afforded by web technologies, including geoweb maps.

REFERENCES

Amoore, L. (2009). Algorithmic War: Everyday Geographies of the War on Terror. *Antipode*, *41*(1), 49–69. https://doi.org/10.1111/j.1467-8330.2008.00655.x.

Barry, A. (2012). Political Situations: Knowledge Controversies in Transnational Governance. *Critical Policy Studies*, *6*(3), 324–336. https://doi.org/10.1080/19460171.2012.699234.

Bittner, C. (2014). Reproduktion sozialräumlicher Differenzierungen in OpenStreetMap.pdf. *Kollaborative Kartographie*. http://www.geographie.nat.uni-erlangen.de/wp-content/uploads/Bittner-2014-Reproduktion-sozialr%C3%A4umlicher-Differenzierungen-in-OpenStreetMap.pdf. Accessed 9 February 2016.

Bittner, C. (2016). OpenStreetMap in Israel and Palestine—'Game Changer' or Reproducer of Contested Cartographies? *Political Geography*. http://www.sciencedirect.com/science/article/pii/S096262981630035X. Accessed 27 December 2016.

Crampton, J. W. (2003). Cartographic Rationality and the Politics of Geosurveillance and Security. *Cartography and Geographic Information Science*, *30*(2), 135–148. https://doi.org/10.1559/152304003100011108.

Ford, H., & Graham, M. (2015). *Semantic Cities: Coded Geopolitics and the Rise of the Semantic Web* (SSRN Scholarly Paper No. ID 2682459). Rochester, NY: Social Science Research Network. https://papers.ssrn.com/abstract=2682459. Accessed 31 October 2018.

Ford, H., & Wajcman, J. (2017). 'Anyone Can Edit', Not Everyone Does: Wikipedia's Infrastructure and the Gender Gap. *Social Studies of Science*, *47*(4), 511–527. https://doi.org/10.1177/0306312717692172.

Fuchs, C. (2011). A Contribution to the Critique of the Political Economy of Google. *Fast Capitalism*, *8*(1), 31–50.

Glasze, G., & Perkins, C. (2015). Social and Political Dimensions of the OpenStreetMap Project: Towards a Critical Geographical Research Agenda. In J. J. Arsanjani, A. Zipf, P. Mooney, & M. Helbich (Eds.), *OpenStreetMap in GIScience* (pp. 143–166). Springer International Publishing. https://doi.org/10.1007/978-3-319-14280-7_8.

Graham, M., Hogan, B., Straumann, R. K., & Medhat, A. (2014). Uneven Geographies of User-Generated Information: Patterns of Increasing Informational Poverty. *Annals of the Association of American Geographers*, *104*(4), 746–764.

Graham, M., & Zook, M. (2013). Augmented Realities and Uneven Geographies: Exploring the Geolinguistic Contours of the Web. *Environment and Planning A: Economy and Space*, *45*(1), 77–99. https://doi.org/10.1068/a44674.

Latour, B. (2004). Whose Cosmos, Which Cosmopolitics? *Common Knowledge*, *10*(3), 450–462.

Lee, M. (2010). A Political Economic Critique of Google Maps and Google Earth. *Information, Communication & Society*, *13*(6), 909–928. https://doi.org/10.1080/13691180903456520.

Le Guin, U. K. (2017). A Non-Euclidean View of California as a Cold Place to Be. In *Dancing at the Edge of the World: Thoughts on Words, Women, Places*. New York: Grove Press.

Leszczynski, A. (2012). Situating the Geoweb in Political Economy. *Progress in Human Geography*, *36*(1), 72–89. https://doi.org/10.1177/030913251141 1231.

Leszczynski, A. (2016). Speculative Futures: Cities, Data, and Governance Beyond Smart Urbanism. *Environment and Planning A*, *48*(9), 1691–1708. /https://doi.org/10.1177/0308518X16651445.

Leuenberger, C., & Schnell, I. (2010). The Politics of Maps: Constructing National Territories in Israel. *Social Studies of Science*, *40*(6), 803–842.

Pursley, S. (2015, June 2). 'Lines Drawn on an Empty Map': Iraq's Borders and the Legend of the Artificial State. *Jadaliyya*. http://www.jadaliyya.com/pages/index/21759/lines-drawn-on-an-empty-map_iraq%E2%80%99s-borders-and-the. Accessed 19 February 2018.

Quiquivix, L. (2014). Art of War, Art of Resistance: Palestinian Counter-Cartography on Google Earth. *Annals of the Association of American Geographers*, *104*(3), 444–459. https://doi.org/10.1080/00045608.2014.892328.

Rose-Redwood, R. (2015). Introduction: The Limits to Deconstructing the Map. *Cartographica: The International Journal for Geographic Information and Geovisualization*, *50*(1), 1–8. https://doi.org/10.3138/carto.50.1.01.

Sadowski, J., & Pasquale, F. A. (2015). *The Spectrum of Control: A Social Theory of the Smart City* (SSRN Scholarly Paper No. ID 2653860). Rochester, NY: Social Science Research Network. https://papers.ssrn.com/abstract=265 3860. Accessed 7 May 2018.

Shaw, J., & Graham, M. (2017). An Informational Right to the City? Code, Content, Control, and the Urbanization of Information. *Antipode* 49: 907–927. https://doi.org/10.1111/anti.12312.

Sieber, R. E., & Haklay, M. (2015). The Epistemology(s) of Volunteered Geographic Information: A Critique. *Geo: Geography and Environment*, *2*(2), 122–136. https://doi.org/10.1002/geo2.10.

Star, S. L. (1990). Power, Technology and the Phenomenology of Conventions: On Being Allergic to Onions. *The Sociological Review*, *38*(S1), 26–56. https://doi.org/10.1111/j.1467-954X.1990.tb03347.x.

Thatcher, J. (2013). Avoiding the Ghetto Through Hope and Fear: An Analysis of Immanent Technology Using Ideal Types. *GeoJournal*, *78*(6), 967–980. https://doi.org/10.1007/s10708-013-9491-0.

Walter, M., & Andersen, C. (2013). *Indigenous Statistics: A Quantitative Research Methodology*. Walnut Creek: Left Coast Press.

Wood, D. (2010a). Mapmaking, Counter-Mapping, and Map Art in Palestine. In *Rethinking the Power of Maps*. Guilford Press. Accessed 1 February 2016.

Wood, D. (2010b). *Rethinking the Power of Maps*. Guilford Press. Accessed 1 February 2016.

Zook, M., & Graham, M. (2007a). Mapping DigiPlace: Geocoded Internet Data and the Representation of Place. *Environment and Planning B: Planning and Design*, *34*(3), 466–482. https://doi.org/10.1068/b3311.

Zook, M., & Graham, M. (2007b). The Creative Reconstruction of the Internet: Google and the Privatization Of Cyberspace and DigiPlace. *Geoforum*, *38*(6), 1322–1343. https://doi.org/10.1016/j.geoforum.2007.05.004.

Appendix: Notes on Methods

In Chapter 2, I briefly describe my research methods as comprising a combination of digital methods and fieldwork. Here, I provide some more details on my sources.

The project is grounded in my experience living and working in Jerusalem. As mentioned at the start of the book, between September 2015 and March 2016, I worked as a mapping assistant for GS. I managed the GIS database of the organisation and helped in the preparation of maps and other graphics. I also supported the running of several mapping workshops, where GJ staff worked with residents to map Palestinian neighbourhoods using GPS units. I returned to Jerusalem for six more months, in 2017, to carry out this research. On that occasion, I conducted another series of mapping workshops, in partnership with Grassroots Jerusalem. Most participants were Palestinian students from Hebrew University and Al-Quds University. Women outnumbered men ten to six. The workshops ran weekly from February to May and involved a mix of group discussions and practical exercises, using both paper maps and mapping software.

To identify the controversies analysed in the book, I have primarily drawn on digital methods, building on the recent literature on controversy analysis in the actor–network-theory vein. After setting the browser to minimise the influence of location and search history, I ran a series of Google Search queries of the type: 'Google Maps AND Jerusalem', 'Waze AND Israel OR Palestine', etc. Each search was repeated in Hebrew and

© The Editor(s) (if applicable) and The Author(s) 2021
V. Carraro, *Jerusalem Online*, The Contemporary City,
https://doi.org/10.1007/978-981-16-3314-0

Arabic. For each query, I considered the top 50 results, looking for references to debates, disagreement or concern around the working of these maps. This method proved effective for revealing debates around digital maps, which often first emerge on user forums, or through the personal blogs of dedicated amateur mappers. A further advantage has been the possibility to access non-English language sources using automated translation software. The translation quality allowed me to understand the general meaning of this material, though of course not to appreciate every nuance. The main drawback of this choice, on the other hand, has been the difficulty of reconciling it with my commitment to investigate neglected issues, starting from the perspective of marginalised actors.

My fieldwork allowed me to partially counter these limitations. A small survey ($n = 70$) conducted both online (through Google Forms) and offline (with paper questionnaires distributed through the mapping workshop participants) helped me identify mapping services that were relevant for local users. I also conducted fifteen semi-structured, one-hour long with map users. At the start, the interviews involved general questions about the respondent's experience with maps in Jerusalem and other places. As the research progressed, they became more focused on the apps and controversies examined in the case study chapters. The interviewees were recruited partially through snowball sampling and partially through the survey, which gave respondents the option to leave their contacts. Among the interviewees, six were Israelis, five were Palestinian and four were visitors from abroad. Men outnumbered women ten to five, and four of the interviewed women were personal acquaintances. Several interviewees had expertise or a personal interest in mapping. These included, for example, a GIS student, a committed OSM mapper, a software developer and a man who self-identified as a map enthusiast. While this small sample is clearly not representative of map users in Jerusalem, the aim of these interviews was exploratory, with no intention to draw general conclusions.

To research Waze and WADA, I drew largely on two types of sources: discussion threads in the Waze user forums, and news articles or blog posts about the controversy. Of the publications discussing Waze in the Israel/Palestine context, 28 appeared on English-language outlets, 20 on Hebrew-language outlets and 7 on Arabic-language outlets. In addition, I examined 60 more articles that considered WADA or similar apps in other places, notably the US and Brazil. The chapter on Google is based on the analysis of the Google Maps interface, of the *Google Public Policy*

Blog content, and of 91 tweets about the Google glitch. I also considered the press coverage of this episode, examining 20 related news articles in Arabic and 20 in English. Among these, four were from Palestinian news outlets. While a search for publications in Hebrew did not yield any results, there are four Israeli publications among the Arabic and English sources. Finally, for the chapter on OSM, I drew on my experience and conversations with GJ, as well as on 'OSM sources', such as the OSM database, the user forum (55 posts by eight users) and mailing list (54 posts by 24 users). The section on the historical development of street names and street signs in Jerusalem is based on secondary sources and on the analysis of historical maps from the Israel National Library's Eran Laor Cartographic Collection.

INDEX

A
Algorithms, 3, 16–18, 25, 28, 45, 47, 55, 57, 68, 75–77, 81, 82, 112–114, 117

B
Big Data, 15, 28, 30, 82
Bordering, 57, 114
British Mandate, 2, 78, 101, 118

C
Critical cartography, 2, 3, 15, 18, 30, 32, 115, 117, 118
Crowdsourcing, 13, 16, 17, 19, 31, 79, 88, 89, 96, 99, 113, 114, 117, 118

D
Danger-tracking apps, 45–49, 54–56, 58

G
Geoweb, 13–19, 34, 75, 88, 96, 99, 111–118, 120
GIS, 106, 125, 126
Google, 1–3, 15, 17, 18, 34, 51, 66–82, 88, 115, 125–127
Grassroots Jerusalem (GJ), 2, 3, 19, 89–95, 105, 125, 127

M
Matters of concern/care, 33, 35, 36

O
OpenStreetMap (OSM), 3, 17, 19, 24, 34, 44, 53, 87–99, 104–106, 113, 115, 117, 120, 126, 127
Oslo Accords, 11, 49

P
Publics, 18, 19, 25, 32–34, 49, 51, 52, 57–59, 71, 77, 78, 80–82, 88, 92, 98, 100, 102, 112, 114, 117, 127

S
Semantic web, 69, 113
Settler colonialism, 56
Shu'fat, 52, 53, 103, 104

T
Twitter, 19, 34, 65, 77, 81, 117

W
Waze/WADA, 17, 18, 34, 44–47,
    49–52, 54–59, 87, 113–115,
    117, 120, 125, 126

Z
Zionism, 4, 5, 9, 28, 120